3 Progress in Clinical Biochemistry and Medicine

Metabolic Control in Diabetes Mellitus
Beta Adrenoceptor Blocking Drugs
NMR Analysis of Cancer Cells
Immunoassay in the Clinical Laboratory
Cyclosporine

With Contributions by
W. Berger, R. Flückiger, H.G. Köppe, K.T. Holmes,
C. E. Mountford, E. L. Nickoloff, T. G. Payne,
M. H. Schreier, I. C. P. Smith, R. M. Wenger

With 68 Figures

Springer-Verlag
Berlin Heidelberg New York Tokyo

As a rule, contributions to this series are specially commissioned. The editors and publishers will, however, always be pleased to receive suggestions and supplementary information. Papers are accepted for "Progress in Clinical Biochemistry and Medicine" in English.

Library of Congress Cataloging-in-Publication Data. Metabolic control in diabetes mellitus. (Progress in clinical biochemistry and medicine ; 3) Includes index. 1. Diabetes. 2. Adrenergic beta blockers. 3. Cancer-Diagnosis. 4. Magnetic resonance imaging. 5. Immunoassay. 6. Cyclosporin A. I. Berger, W. (Willi), 1933–. II. Title: Metabolic control in diabetes mellitus. III. Series. [DNLM: 1. Adrenergic Beta Receptor Blockaders. 2. Cyclosporins-pharmacodynamics. 3. Diabetes Mellitus-metabolism. 4. Immunoassay. 5. Neoplasms-analysis. 6. Nuclear Magnetic Resonance
W1 PR668EM v. 3 WK 810 M587]
RC660.M47 1986 616.4'62 86-1787

ISBN-13: 978-3-642-71000-1 e-ISBN-13: 978-3-642-70998-2
DOI: 10.1007/978-3-642-70998-2

Softcover reprint of the hardcover 1st edition 1986

2152/3140-543210

Editorial Bord

Table of Contents

Monitoring of Metabolic Control in Diabetes Mellitus: Methodological and Clinical Aspects

Rudolf Flückiger and Willi Berger

Departments of Research and Internal Medicine
University Clinics Basel, CH-4031 Basel, Switzerland

This is a review of the major advancements in monitoring techniques and strategies used in the management of diabetes mellitus. The major contributions comprise

1) the reagent strip methodology for self-monitoring of glucose in blood and urine,
2) the measurement of hemoglobin glycosylation for the quantitative assessment of long-term glycemia,
3) the C-peptide assay for the evaluation of residual pancreatic function, and
4) the development of clinical laboratory methods to detect an early abnormality of renal function.

The information provided outlines the principles, scope, and limitations of these methods in order to allow optimal use of these new possibilities.

Progress in Clinical Biochemistry
and Medicine, Vol. 3

1 Introduction

The close association between blood glucose control and the well-being of the patient, as well as the risk for the development of the "late" complications of diabetes, make it necessary to attain near normalisation of blood glucose[1-3]. Significant progress has been made in this direction in the last few years because of the advancement of analytical techniques for the monitoring of both metabolic status and the functional state of the pancreas and the kidneys, organs involved in the disease process. The respective methodologies are the test strips for self-monitoring of glucose in blood and urine, the measurement of the nonenzymatic glycosylation of hemoglobin and serum proteins, the C-peptide assay, and the determination of small amounts of albumin in urine.

The *test strip methodology* for determination of glucose in blood and urine has made possible home blood glucose monitoring which enables the patient to aim for treatment targets near the physiological range. To this information on short term glycemia obtained by the patient, the determination of *hemoglobin glycosylation* in the clinical chemistry laboratory adds quantitative information about averaged long-term glycemic control.

The *C-peptide assay* allows evaluation of the residual pancreatic function in the presence of exogenous insulin. Results of a C-peptide assay are helpful in selecting the appropriate treatment for poorly controlled maturity-onset diabetes.

The detection of *microalbuminuria,* an abnormal albumin excretion below the level of "Albustix" detection, establishes nephropathy before renal damage becomes irreversible.

2 Assessment of Glycemia in Type I and Type II Diabetes

Fluctuations of blood glucose concentration are fundamentally different in insulin-dependent (type I, IDDM) and non-insulin-dependent diabetes mellitus (type II, NIDDM) (Fig. 1) and require different monitoring strategies.

In the insulin-dependent type I diabetes, blood glucose concentrations fluctuate widely thus necessitating regular glucose monitoring in blood or urine in order to adjust the insulin dosage and injection schedule to the daily needs of the body. The blood glucose concentration measured at any time is largely determined by the dosage and composition of the insulin administered as well as diet. The fasting blood glucose concentration depends on the slow-acting insulin dose of the previous evening, that determined before dinner on the slow-acting insulin administered in the morning, and the blood glucose concentration determined before lunch or at bedtime on the fast-acting insulin dose of the morning or evening respectively. Postprandial blood glucose concentrations depend on the diet and the action of the fast-acting insulin.

Hypoglycemia is the main adverse side effect of insulin therapy. The probability of hypoglycemic episodes increases with attempts to attain near-normoglycemia

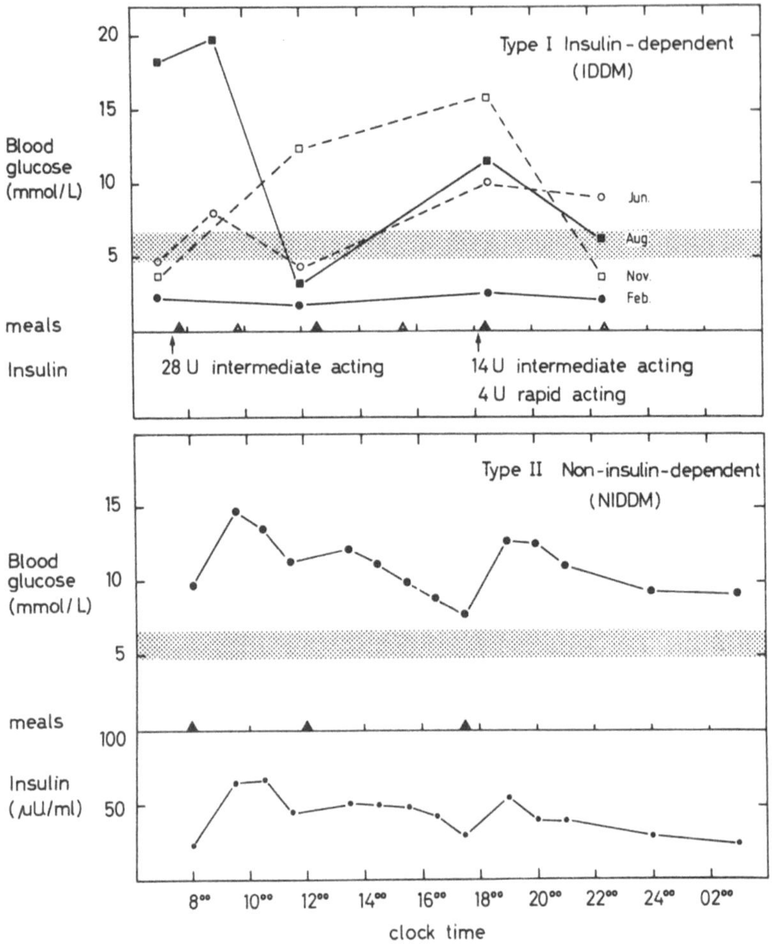

Fig. 1. Fluctuations of blood glucose concentrations in insulin-dependent and non-insulin-dependent diabetic patients. ░ reference range; ▲ meals or snacks

using more aggressive therapy. A special risk of hypoglycemia exists at night when its occurrence is not recognized. Since such episodes tend to occur between 2 and 4 a.m., a blood glucose determination at this time would enable detection of nocturnal hypoglycemia. The risk of nocturnal hypoglycemic episodes can be estimated from the blood glucose concentration before bed, being 50% if the blood glucose is below 5.5 mmol/l[4].

In the non-insulin-dependent, type II diabetes, blood glucose concentrations fluctuate in a predictable manner. Impairment of insulin secretion in type II diabetic patients is delayed and protracted in the morning hours resulting in poor glucose utilization after breakfast but better utilization during and after lunch. Thus, blood glucose typically increases by 5–6 mmol/l after a breakfast containing 40–50 g carbohydrate, while this increase is only 2–4 mmol/l after an isocaloric lunch containing the same amount of carbohydrate. The lowest blood glucose concentration in type II

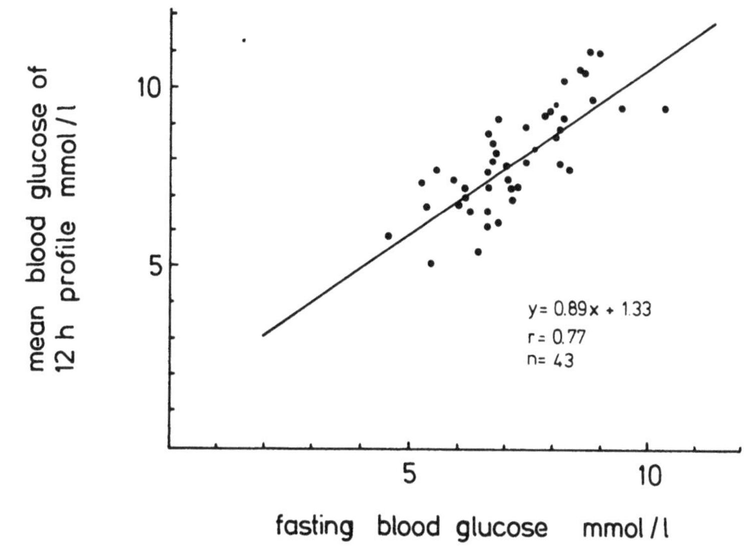

Fig. 2. Correlation between fasting blood glucose and mean glycemia in type II diabetic patients

diabetics is approached 3 to 5 h after lunch or late in the evening. These blood glucose patterns remain remarkably constant from day to day[5].

In patients on sulfonylurea therapy, the decrease in blood glucose is more pronounced than in patients treated by diet alone. In such patients "after-lunch" values may be 25% lower than fasting blood glucose concentrations[5]. It is important to be aware of this peculiarity of blood glucose variations in sulfonylurea-treated type II diabetics because a normal or near-normal blood glucose value will frequently be associated with an abnormally low afternoon blood glucose concentration. For this reason, determination of the fasting blood glucose concentration alone is not sufficient to asseses the risk of hypoglycemia in type II diabetics on sulfonylurea therapy.

This risk of hypoglycemic episodes is best assessed by the blood glucose concentration 3–5 h after lunch. The fasting blood glucose, on the other hand, reliably reflects the mean glycemia in type II diabetics[5] (Fig. 2). In some patients accentuated diurnal variations in blood glucose result in a low afternoon blood glucose concentration and a high fasting blood glucose concentration thus erroneously suggesting poor metabolic control.

3 Methodology for the Determination of Glucose

Much effort has been invested in the search for simple and specific methods for the determination of glucose in biological fluids (for a review see[6]). The principal approaches are summarized in Table 1 and fall into two categories, namely early chemical methods and recent enzymatic methods.

Table 1. The principal approaches for the determination of glucose

1. *Reducing methods*
The highly reactive enediol of glucose is oxidized by numerous oxidants and measured as below:

oxidant /	reduced product	measurement	designation
picric acid /	picramic acid	– red picramic salt	Crecelius-Seifert
ferricyanide /	ferrocyanide	– iodometric titration of reduced product	Hagedorn-Jensen
		– oxidant consumtion by photometry	Hoffmann ferricy-anide
copper(II) /	copper(I)	– red cuprous oxide	
– tartrate			Fehling's solution
– citrate			Benedict reagent in Clinitest
		– "molybdenum blue" formed by reduction of phosphomolybdic acid by cuprous oxide	Nelson-Somogyi
		– neocuproine complex of copper(I)	Folin-Wu
silver(I) /	elemental silver		Tollen's reagent

2. *Hydroxymethylfurfural – phenol condensation methods*	
Dehydration of glucose in hot acid yields hydroxymethylfurfural which forms colored phenol adducts	resorcinol test anthrone
3. *Aldose-aromatic amine condensation methods*	
Glucose condenses with aromatic amines to produce colored glycosyl-amines	o-toluidine
4. *Glucose oxidase/peroxidase method*	GOD/POD
Glucose is oxidized by glucose oxidase to gluconolactone and hydrogen peroxide (which is decomposed by peroxidase). Oxygen is measured	
– with an oxygen electrode	Beckmann analyzer
– by reaction with chromogenic oxygen acceptors:	
o-anisidine	in Dextrostix
o-tolidine	in Clinistix or Tes Tape
5. *Hexokinase/glucose-6-phosphate dehydrogenase method*	
Hexokinase converts glucose to glucose-6-phosphate. The dehydro-genase generates 6-phosphogluconate and NADPH (which is measured spectrophotometrically)	

The specificity of methods for the determination of glucose is usually assessed in terms of quantitative agreement with values obtained by enzymatic methods. Although the latter values are frequently referred to as "true" glucose, the presence of inhibitors can influence the accuracy of the enzymatic methods. Of the chemical methods, the o-toluidine method is least affected by interfering substances. In the methodology which is based on the reducing properties of glucose, interference arises from nonglucose reducing substances, termed the "saccharoid" fraction, which includes glutathione, creatinine, uric acid, and ascorbic acid.

When compared with the enzymatic methods, results of the ferricyanide reaction and the Somogyi-Nelson method are ≤7% higher[7, 8], while those obtained with the o-toluidine method are virtually identical.

Today, most clinical laboratories use enzymatic methods based on hexokinase or glucose dehydrogenase. These methods allow precise measurement of glucose concentrations with coefficients of variation of the order of 5%.

3.1 Interpretation of Blood Glucose Values

The basic parameter of metabolic control in diabetes is the concentration of glucose in the blood. Blood glucose measurements are performed on whole blood, plasma, or deproteinized plasma. The difference in respective values is large enough to necessitate adjustment of critical values for screening and diagnostic purposes.

At normal hematocrits of 0.4–0.5, plasma glucose values are 13–15% higher than those determined on whole blood. Deproteinization of plasma increases glucose values by 5%. The lower whole blood glucose concentration is a consequence of the lower water content of the erythrocyte (65%) compared with plasma (93%) and the identical glucose concentration in these two compartments[9]. The relationship between whole blood glucose (WBG) and plasma glucose (PG) concentration depends on the hematocrit (hct) as

$$\text{WBG} = \text{PG}\,(1 - 0.3\,\text{hct}) \tag{1}$$

allowing conversion even in conditions of grossly abnormal hematocrits. For clinical use:

$$\text{PG} = 1.15 \times \text{WBG}\ . \tag{2}$$

Sample collection is an important consideration in blood glucose determinations. *In vitro* glycolytic degradation of glucose in whole blood is quite variable and averages 7%/h. To avoid glycolysis either plasma or serum is separated within 30 min or an inhibitor of glycolysis is added. The inhibitors sodium fluoride or ammonium fluoride (0.9 mg fluoride/mg blood) are used, the latter if sodium determination is required on the same sample[10]. Serum removed immediately after separation of whole blood yields reliable values for up to 48 h[11].

The glucose concentration in capillary blood is generally higher than in venous blood. This reflects the arteriovenous difference generated by extrahepatic glucose utilization. According to the WHO the arteriovenous difference between capillary and venous blood specimens is 1.1 mmol/l (= 20 mg%), except in the case of fasting blood glucose values[12].

Interconversion of “blood glucose” parameters from the popular weight/volume dimensions to mmol/l (SI-units) is recommended and requires multiplication of the mg% values by 0.0555.

3.1.1 Diagnosis of Diabetes

Opinion on the diagnostic criteria of diabetes is under continuous revision. Glycosuria, fasting blood glucose, and the response to a glucose challenge have all been advocated. Fasting blood glucose values are again now more widely used than oral glucose tolerance tests.

In normoglycemic persons the fasting plasma glucose concentration is usually < 6.4 mmol/l (< 115 mg%). Its value increases somewhat with age and is subject to seasonal variation[13, 14]. A fasting venous plasma glucose concentration > 7.8 mmol/l (> 140 mg%) was arbitrarily defined by the Diabetes Data Group as abnormal or diabetic[15]. If fasting venous plasma glucose values > 6.4 mmol/l but < 7.8 mmol/l are observed, further evaluation of the glucose intolerance by glucose tolerance tests is recommended. The criteria during pregnancy are more stringent. Here, the upper limit of a normal whole fasting blood glucose is 5.0 mmol/l (90 mg%).

Oral glucose tolerance tests are influenced by a variety of factors and in order to obtain reproducible results the performance of the test must be strictly standardized. Food intake should contain at least 1500 Calories/day and normal physical activity is recommended for 3 days prior to the test. Complete physical inactivity results in some glucose intolerance. Certain types of medication such as thiazide-diuretic, beta-blocker, or salicylate preparations should be discontinued. The test should be performed after an overnight fast of 10–14 h. Water is permitted but smoking is not. The procedure is as follows: first a fasting blood sample is taken after which the sitting patient drinks 75 g of glucose in 250–350 ml water within 5–15 min. Blood samples are taken 2 and sometimes 3 h thereafter. Two-hour venous plasma glucose concentrations reaching 11 mmol/l (200 mg%) must be considered as a diagnosis of diabetes for patients of all ages, according to the NIH Diabetes Data Group. Based

Table 2. Diagnostic values for oral glucose tolerance tests under standard conditions. Load 75 g glucose in 250–350 ml water for adults or 1.75 g/kg body weight (to a maximum of 75 g) for children, using specific enzymatic glucose assay

		Glucose concentration		
		Venous whole blood	Capillary whole blood	Venous plasma
Diabetes Mellitus				
Fasting	**mmol/l**	**≥ 6.7**	**≥ 6.7**	**≥ 7.8**
	mg%	≥ 120	≥ 120	≥ 140
and/or				
2 h after glucose	**mmol/l**	**≥ 10.0**	**≥ 11.0**	**≥ 11.0**
load	mg%	≥ 180	≥ 200	≥ 200
Impaired glucose tolerance				
Fasting	**mmol/l**	**< 6.7**	**< 6.7**	**< 7.8**
	(mg%)	< 120	< 120	< 140
and				
2 h after glucose	**mmol/l**	**≥ 6.7–< 10.0**	**≥ 7.8–< 11.0**	**≥ 7.8–< 11.0**
load	(mg%)	≥ 120–< 180	≥ 140–< 200	≥ 140–< 200

Table 3. Criteria for the diagnosis of gestational-onset diabetes

	Two or more of the following values after a 100 gm oral glucose challenge must be met or exceeded:			
	Venous whole blood		Venous plasma	
	mmol/l	mg%	**mmol/l**	mg%
Fasting	**5.0**	90	**5.8**	105
One-Hour	**9.5**	170	**10.6**	190
Two-Hour	**8.1**	145	**9.2**	165
Three-Hour	**7.0**	125	**8.1**	145

on these recommendations the WHO expert committee on diabetes mellitus proposed the diagnostic criteria in Table 2. Long-term evaluation of glucose tolerance during pregnancy has been performed by O'Sullivan and Mahan with 100 gm oral glucose tolerance tests[16)] which yielded the criteria for gestational diabetes given in Table 3.

Intravenous glucose tolerance tests are also used for the diagnosis of diabetes. These tests yield more reproducible results than do oral glucose tolerance tests because they are not affected by gastrointestinal absorption. In the most commonly used protocol a bolus of glucose (0.33 g/kg but not more than 25 gm as 50 ml of a 50% solution) is infused into a deep brachial vein within 2 min. After 10 min to allow for equilibration with the extracellular space, samples are drawn at approximately 5-minute intervals over a 30–45 min period. Results are plotted semilogarith-

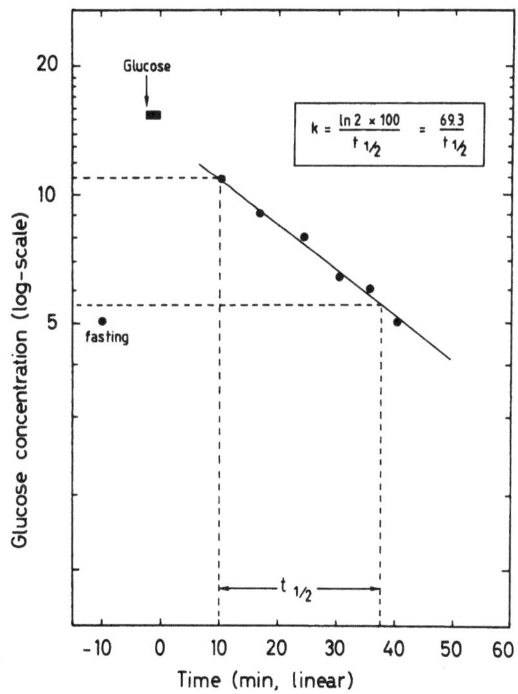

Fig. 3. Evaluation of intravenous glucose tolerance test

mically and the line of best fit is drawn. The blood glucose decay constant k is computed by the formula shown in Fig. 3. Values of k below 1.0 are abnormal and values above 1.2 are normal[17)].

3.1.2 Reagent Strip Methodology

Reagent strips consist of paper impregnated with a glucose-peroxidase-chromogen system. The indicator is oxidized by glucose, and the color produced and its intensity are related to the glucose concentration. In the process of exposure, the plasma penetrates the indicator field, and test strips therefore measure plasma glucose rather than whole blood glucose despite the application of whole blood to the strips[18)].

Two test strip types differing in the number of color blocks are currently in use: the Dextrostix (Reflocheck-Glucose) with one color block and the Chemstrip (Haemo-Glukotest 20–800 R) with dual color blocks[19)]. Accuracy obtainable with the visually read Chemstrip is comparable to that obtained by reading Dextrostix strips with a reflectance meter. In addition, color stability is better with the Chemstrip technique which makes accurate timing less critical[20)]. Overexposure and underexposure yield overestimates and underestimates, respectively, of blood glucose concentration. All the reagent strips available for self-monitoring of blood glucose are sufficiently accurate, provided the tests are performed by properly instructed persons[20, 21)].

Accuracy is best in the practically important range of 5–10 mmol/l with a trend for greater deviations at higher glucose values. A deviation of a blood glucose estimate of up to 20% from the actual value is acceptable since this is unlikely to lead to inappropriate adjustments of insulin therapy. More precise readings are obtained by use of a reflectance meter. Use of a reflectance meter is also indicated for diabetics with visual impairment[22)].

3.1.3 Assessment of Variability of Glycemia: Blood Glucose Profiles

Daily blood glucose profiles can now easily be obtained by the patient at home. Such profiles are of utmost importance for the adjustment of the insulin dose in insulin-dependent diabetics. Figure 4 shows a 48-hour sampling profile. It can be seen that adequate information about the stability of glycemic control is obtained only if all preprandial values, the postprandial value after breakfast, and the bedtime and the nocturnal value are obtained. More frequent sampling generally does not improve information about the variability of glycemia.

The commonly used sampling at breakfast, lunch, and supper clearly does not adequately reflect the daily fluctuations of glycemia in labile diabetic patients.

Variability of glycemia is often expressed numerically by the M-value[23)]. The M-value is a quantitative index of the deviations of several blood glucose (BG) determinations in a 24-hour period from an arbitrarily selected standard (originally 120 mg%). For enzymatic blood glucose measurements the relevant standard value is 80–100 mg%. The formula was designed to give proportionally greater emphasis

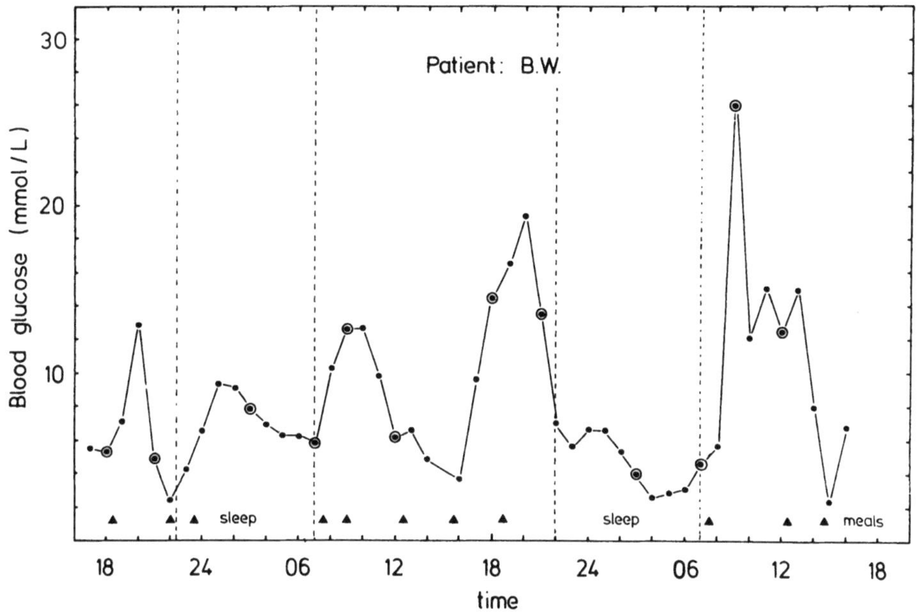

Fig. 4. A 48-hour sampling blood glucose profile in a type I diabetic patient. ⊙ preferred sampling hours

to hyperglycemia. The 24-hour M-value is calculated as the mean, M (BG/BG), calculated for the various blood glucose values according to formula (3):

$$M\left(\frac{BG}{BG}\right) = 10\left(\log\frac{BG}{120}\right)^3 \tag{3}$$

The M-value is obtained by adding the term M(W) = W/20, for fluctuations in which W is the difference between the maximum and minimum blood glucose values during the period concerned. This yields formula (4):

$$M = M(BG) + M(W) \tag{4}$$

in which M(BG) is the average of all M(BG/BG) values [formula (3)]. M-values ranging from 0–18, 19–31 and 32–, are obtained for patients with good, fair, and poor glycemic control, respectively.

The MAGE value, the mean amplitude of glycemic excursion, is an alternative measure often used to determine stability of glycemia. The MAGE value is calculated by taking the arithmetic mean of the blood glucose increases or decreases (from blood glucose nadirs to peaks or vice versa) when ascending or descending segments exceed the value of one standard deviation of the blood glucose for the same 24-hour period. (For details of calculation compare the Appendix of[24].) Normal MAGE values are 22–60 mg/100 ml, with 67–82 mg/100 ml for stable diabetics and 119–200 mg/100 ml for unstable diabetics.

3.2 Urinary Glucose Monitoring

Measurement of the degree of glycosuria forms the basis for home monitoring, particularly in type II diabetic individuals. The popularity of urinary glucose determinations has declined since the introduction of the simple determination of blood glucose and because of the discrepancies often found between blood and urinary glucose measurements[25]. Nevertheless, urinary glucose measurements have not lost their importance. Urinary glucose measurements provide a very useful though rough guide to blood glucose control and distinguish mainly between blood glucose values above and below the renal threshold[26]. In contrast to that for monitoring of blood glucose, sample collection for urinary glucose monitoring is not painful. Furthermore, this determination does not involve special skill or experience as does the determination of blood glucose.

3.2.1 Methods for Self-Monitoring of Urinary Glucose

The reducing methodology (see Table 1) is the basis of the convenient Clinitest tablets (Ames, Merck) for self monitoring of urinary glucose. The availability of test strips for urine glucose determinations has decreased the use of the Clinitest method. Although this determination requires more manipulation than the test strips it has an advantage in that different glucose concentrations result in different colors (i.e. blue: no glucose, brown: 1–2% glucose, red: more than 2% glucose).

There are various enzyme-based test strips with differing sensitivities and ranges. Tes-Tape (Lilly) or Glukotest (Boehringer, Mannheim) strips are well suited for merely qualitative information while the Diabur 5000 (Boehringer, Mannheim) test yields semi-quantitative information. Diabur 5000 gives accurate measurements of urinary glucose up to a concentration of 5% by means of a combined bicolored test field. Within a reaction time of 2 min, glucose concentration steps of – / 0.1 / 0.25 / 0.5 / 1 / 2 / 3 / 5% can be reliably differentiated. In addition, interference by other urine constituents, especially ascorbic acid and ketone bodies, is largely eliminated[27,28].

The two most important urinary factors causing misleading results with the Clinitest tablets and/or Tes-Tape are ascorbic acid and gentisic acid, a metabolite of salycilates[29]. These reducing agents cause erroneously high values with Clinitest which is based on reducing methodology and low values with the glucose oxidase tests (Tes-Tape) by keeping the indicator dye in its reduced form[29,30].

Interpretation of results from urinary glucose determinations requires consideration of the sampling characteristics and the renal threshold for glucose.

Glucose appears in the urine in significant amounts only if the blood glucose concentration in blood exceeds its renal threshold. In young persons the renal threshold for glucose lies between 7.2–10.0 mmol/l (130–180 mg%) and gradually increases with age to reach values of 11.1–13.9 mmol/l (200–250 mg%)[31,32]. During pregnancy, its value decreases because of increased glomerular filtration[33].

Although the variability of the renal threshold for glucose is pronounced, it remains rather constant over several months in each individual[31,32]. It can easily be estimated on an outpatient basis by comparison of blood glucose and urine glucose profiles[31]. Should an accurate value be required, the fractional excretion rate can be determined according to formula (5):

$$\text{fractional excretion rate} = \frac{\text{urinary glucose} \times \text{plasma creatinine}}{\text{urinary creatinine} \times \text{plasma glucose}} \quad (5)$$

In normal control persons the mean value of the fractional excretion rate is 4.7×10^{-4} with 95% tolerance limits of 2.9 and 9.3×10^{-4} [34].

The clinical implications of the variability of the renal threshold can be estimated from the data of Walford[31]. In this study, the mean renal threshold of the diabetic patients (of mean age 35), was found to average 130 mg% which is comparable to that of nondiabetics. Half of the patients with blood glucose values below 135 mg% were glycosuric and in patients with 2% glycosuria blood glucose levels varied from 144–360 mg%. Therefore, heavy glycosuria (≥ 2%) can result in an overestimation of the blood glucose concentration.

The determination of glucose in urine reflects the acute blood glucose concentration only if the bladder is emptied and the determination performed on urine collected during the subsequent 30–60 min ("double voided"). Figure 5 provides an example of the relationship between blood glucose and results of urinary glucose measurements after both short and long collection periods.

In insulin-dependent diabetics, urinary glucose determinations are usually performed on "double voided" urine collected prior to the meals and in the evening before bed. Results are recorded in a manner to allow the recognition of trends in metabolic control at a glance. The quantitative determination of urinary glucose in urine collected during 12- or 24-hour periods also needs to be performed at regular time intervals. Some patients with a low renal threshold may excrete 100–200 g glucose per day resulting in a caloric deficit and therefore ketonemia and ketonuria. This type of ketonuria has to be differentiated from that resulting from severe metabolic decompensation.

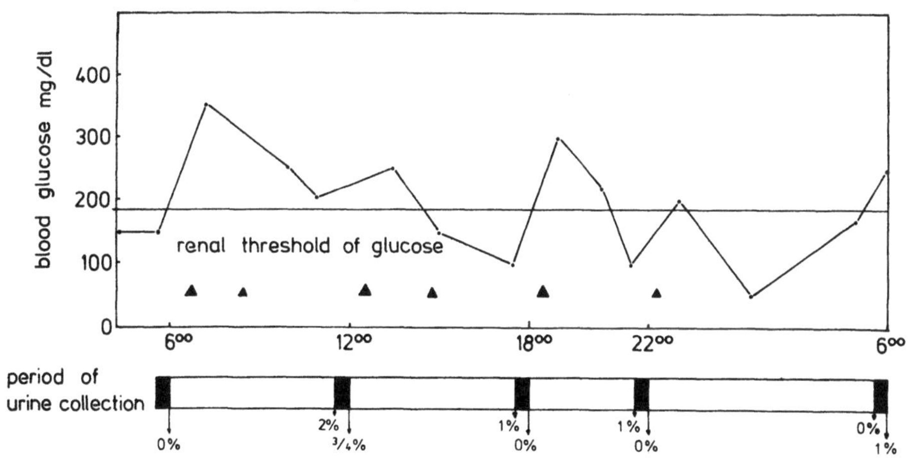

Fig. 5. Relation between blood glucose profile and urinary glucose concentration after short and long urine collection periods

4 Assessment of Long-Term Glycemia: Nonenzymatic Protein Glycosylation

4.1 Chemistry and Methodology

Nonenzymatic protein glycosylation is increased in diabetes in proportion to previous glycemia. The glycosylation of hemoglobin has proved useful for monitoring purposes whereas the glycosylation of many other proteins exposed to increased glucose concentrations has been alleged to be an event responsible for the development of some of the late complications of diabetes.

The chemistry of this reaction is briefly the following (Fig. 6) (review[35]): In the initial step the carbonyl group of glucose reacts with an amino group on the protein to form a Schiff's base or aldimine. This first step is rapid and reversible. Subsequently the aldimine may rearrange to the stable ketoamine (Amadori rearrangement) in a slow reaction. The resulting glycosylated products are fructosyl amino acids. These compounds are not hydrolyzed under strongly acidic conditions, which renders the *specific detection* of nonenzymatic glycosylation difficult.

Quantitation of overall protein glycosylation has been most often performed by a colorimetric test which measures hydroxymethylfurfural released from the nonenzymatically glycosylated proteins under mildly acidic conditions. Hydroxymethylfurfural is detected with thiobarbituric acid as the colored condensation product (TBA-test) (for a review see[36]). Recently, attractive alternatives have become available, like the furosine assay[37], boronate affinity chromatography[38], and the fructosamine assay[39].

Indirect estimation of the extent of nonenzymatic protein glycosylation relies on the fact that the occurrence of this modification reaction under certain circumstances decreases the positive charge of a protein. Alterations of the charge properties by processes other than the nonenzymatic glycosylation yield erroneous estimates of protein glycosylation if charge-dependent techniques are used (see below).

4.2 Glycosylated Hemoglobin

Glucose reacts with numerous amino groups of the hemoglobin molecule while the phosphorylated intermediates of glycolysis, glucose-6-phosphate and fructose-1,6-diphosphate react only with the N-terminal amino group of the beta chains.

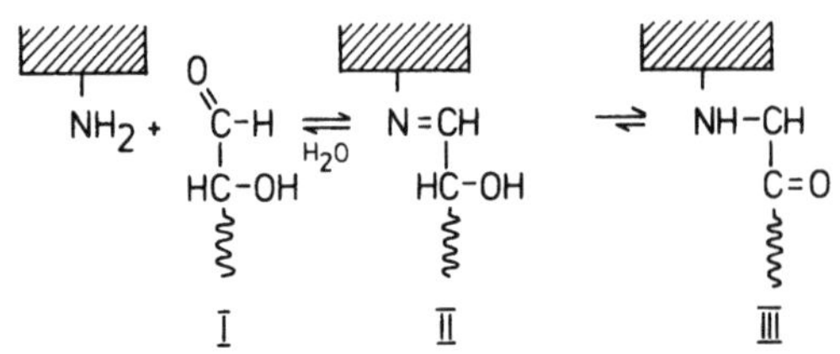

Fig. 6. Chemistry of the nonenzymatic glycosylation reaction:
I glucose; II Schiff base; III ketoamine

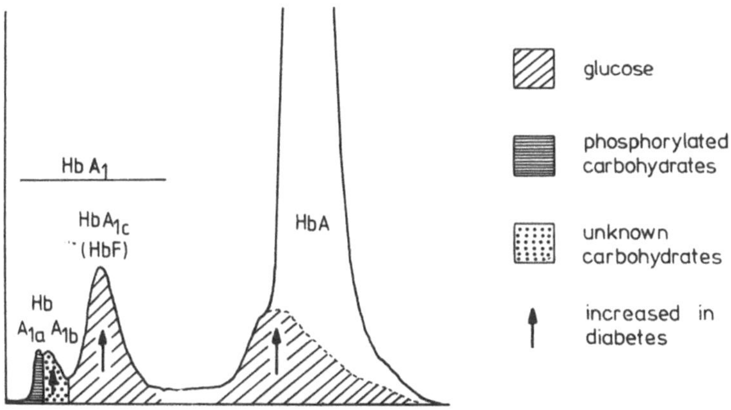

Fig. 7. Schematic representation of charge-dependent separation of a hemolysate: Nomenclature of hemoglobin components and identity of bound carbohydrate moieties

Glycosylation at the N-terminus alters the charge of hemoglobin sufficiently to allow separation of the glycosylated products by charge-dependent techniques. Figure 7 illustrates such a separation and also indicates the nomenclature of the glycosylated hemoglobin components. Three principal minor hemoglobin components are resolved from the main HbA fraction. These minor components – HbA1a, HbA1b, and HbA1c – are collectively referred to as the HbA1 fraction and represent approximately 7% of total hemoglobin in nondiabetics and 2–3 times this amount in diabetic patients. The component HbA1c contains glucose, HbA1a glucose-6-phosphate and fructose-1,6-diphosphate, and HbA1b an as yet unidentified carbohydrate. An amount of glycosylated hemoglobin corresponding to that of HbA1c is not contained in the HbA1 fraction. In diabetes, the concentration of HbA1c, HbA1b, and the unresolved fraction are increased. Note that, unless special analytical efforts are made, fetal hemoglobin, HbF, and the labile HbA1c, the adduct of hemoglobin and glucose, coelute with HbA1c.

4.2.1 Methodological Considerations

A wide variety of methods for the determination of glycosylated hemoglobins based on charge separations have been developed including cation exchange chromatography performed in minicolumns or as high pressure liquid chromatography, isoelectric focusing, or gel permeation chromatography. Of these methods, cation exchange chromatography is most widely used (reviews[40, 41]).

The TBA-reaction is so far the only methodology for the specific quantitation of glycosylated hemoglobin that has been used in clinical routine. A promising immunological approach has been described[42], however, the antibody has not become generally available. Boronate affinity chromatography is another recent addition to the specific methodologies. However, results obtained with this technique tend to be quite variable and have proved to be difficult to control.

Table 4. Conversion of HbA1c and HbA1 values[a]

Glycemic control	HbA1c +	HbA1a + b =	HbA1
Normoglycemia	5%	2%	7%
Moderate hyperglycemia	**7.5%**	**2.5%**	**10%**
Pronounced hyperglycemia	10%	3%	13%
Extreme hyperglycemia	15%	4%	19%

[a] The equation for the relationship is[43]: HbA1a + b = 0.2 HbA1c + 1.18

HbA1c vs. HbA1. The determination of the HbA1 fraction is technically simpler than that of HbA1c. Hence, HbA1 is often determined and interpretation of results requires interconversion of these parameters. Because of the linear relationship between HbA1a + b and HbA1c, interconversion is straightforward and easily memorized (Table 4). During storage of hemolysates for short times HbA1a + b levels tend to increase while the concentration of HbA1c remains constant for up to a month in samples stored as whole blood or hemolysates at 4 °C[44].

Labile HbA1c is formed rapidly in proportion to the acute blood glucose concentration and can account for over 10% of HbA1c. As a consequence HbA1c yields reliable information about long-term blood glycemia only if labile HbA1c is removed. Labile HbA1c can be removed during sample preparation by performing hemolysis at pH 5[45].

The importance of removing *labile HbA1c* is demonstrated by data reported by Goldstein[46]. In this experiment total and stable HbA1c was determined in diabetics and control persons fasting and also 2 and 5 h after a standard breakfast. In the control subjects, the amount of labile HbA1c was small while in the diabetic patients it was elevated two hours after breakfast and constituted more than 10% of the total HbA1c. The resulting misinterpretation of the average long-term glycemia is 60 mg%, which is not acceptable.

Interference by hemoglobin with altered charge

1. *HbF:* Interference by HbF has to be considered in the interpretation of HbA1c or HbA1 measurements during pregnancy. During a normal pregnancy, the concentration of HbA1c or HbA1 remains unchanged or may even decrease slightly in the 2nd trimester[47]. This behavior is the result of at least two opposite trends: a decrease of HbA1c and an increase of the interfering HbF. HbF concentrations may increase during pregnancy to peak in the 24th week and then return to baseline again[48]. On the other hand, glucose tolerance improves until approximately midterm causing HbA1c to decrease. Altered red cell kinetics may accentuate this decrease. Depending on the magnitude of these processes, the "HbA1c" peak remains unaltered, increases, or decreases. The normal range for HbA1c values is 4–6% and for HbA1 6–8%. This reference range is not dependent on age and sex. In females the true HbA1c would probably be about half a percentage point lower because of menstrual blood loss. This difference is often not observed because the HbF is elevated in females and coelutes with HbA1c[44].
2. *Variant hemoglobin:* Measurements of HbA1 and HbA1c do not reliably reflect averaged glycemia if hemoglobin with a charge different from that of HbA is

Table 5. Limitations of HbA1c, HbA1 and glycosylated hemoglobin measurements

Condition	Effect (Magnitude)
Anemias	decreased values
Increased fetal hemoglobin	HbA1c peak increased
Variant hemoglobins	HbA1c/HbA1 decreased
Uremia	HbA1 increased by carbamylated Hb (2%)
Alcoholism	HbA1 increased by modified Hb coeluting with HbA1c (≤2%)
Continuous and highly dosed aspirin or beta-lactam therapy	modified Hb increases HbA1 (≤2%)

present, as is the case with HbF. All of the more common variant hemoglobins, S, C, and D, are more positively charged than HbA and their glycosylated forms elute later from the column with the HbA peak resulting in an underestimation of mean glycemia[49)].

3. *Other modification reactions:* The occurrence of modification reactions other than nonenzymatic glycosylation interferes with the estimation of mean glycemia from HbA1c or HbA1 determinations. Such modification reactions occur in uremia[50)], alcoholism[51)], and in highly dosed aspirin or beta-lactam therapy[52)]. In these conditions specific estimation of hemoglobin glycosylation by the TBA-reaction or boronate affinity chromatography is indicated.

Red blood cell life span. Whenever the red blood cell life span is shortened, the extent of hemoglobin glycosylation does not reliably reflect averaged glycemia[53)]. Table 5 summarizes the more common limitations of HbA1c, HbA1, and glycosylated hemoglobin measurements. In the rare lead poisoning the HbA1 fraction is also increased and increased HbA1 values can result if hyperlipemic blood samples are applied because of interference by lactescence[54)].

4.2.2 Interpretation and Clinical Application

The relationship between averaged glycemia and HbA1c as documented by various studies shows close agreement and allows quantitative interpretation of results (Table 6). In a first approximation an increase of HbA1c by 1 percentage point corresponds to an increase of mean glycemic control by 30 mg%. The above relationship holds only if peak integration or peak collection in the case of minicolumns is ideal. This is not always the case and commercial minicolumns tend to yield values up to 20% lower than those in Table 6.

Glycosylated hemoglobin reflects the averaged glycemia of the previous one to two months. Therefore, its determination every two months reliably documents the mean blood glucose value in patients with type I diabetes. A HbA1c or HbA1 value corresponding to normoglycemia generally cannot be attained in labile type I diabe-

Table 6. Relationship between mean blood glucose and HbA1c

HbA1c (%)	5	6	7	8	9	10	11	12	13	14	Ref.
Blood	**4.2**	**5.9**	**7.8**	**9.7**	**11.9**	**14.2**	**16.4**	**19.1**	**21.2**	**24.9**	(55)
glucose	76	106	140	175	215	255	295	345	396	450	
mmol/l,	**4.5**	**6.3**	**8.2**	**9.9**	**11.9**	**13.7**	**15.5**	**17.4**	**19.2**	**21.1**	(56)
mg%	81	114	147	180	214	247	280	314	347	380	
	4.1	**5.9**	**7.7**	**9.5**	**11.3**	**13.2**	**14.9**	**16.8**	**18.6**	**20.4**	(57)
	74	106	139	172	204	237	270	302	335	368	

tic patients without the risk of frequent and severe hypoglycemic episodes. If near-normoglycemia has to be reached, as is the case during the diabetic pregnancy, this may require 5–8 daily insulin injections or the use of an insulin pump.

A desirable and realistic target value for adolescent type I diabetics has been proposed to be less than 9%[46]. Our experience shows that in adult type I diabetic patients the upper desirable limit can be lowered to 7.5%, the target range for HbA1c being 6.5–7.5% (HbA1 9–10%) for labile type I diabetics. If HbA1c values increase to values above 9% (HbA1 >12%) the treatment is reconsidered. If HbA1c values fall below 6.5% (HbA1 <9%) additional blood glucose measurements are performed to disclose the occurrence of hypoglycemic episodes.

Two examples will illustrate the use of glycosylated hemoglobin measurements in daily clinical practice.

A nurse, 21 years old, developed type I diabetes 2 years ago. At first insulin therapy was easily established with 24 and later 16 units of insulin. With a single daily insulin injection HbA1c values in the range of 6.4–6.8% were achieved. One and a half years after the onset of diabetes, HbA1c values of 11 and 12% were measured. Self-monitoring of glycosuria showed urine to be free of glucose in over 90% of all tests performed. The patient asserted that she felt well and was fit for work. An ad hoc blood glucose determination gave a value of 250 mg% which she explained by a possible mistake in dietary planning. Because we could not explain the discrepancy between the negative urine tests and the high HbA1c value we questioned the correctness of the urine test results. At this point the nurse admitted having manipulated the results because she wanted to avoid to be advised to inject insulin twice a day, a situation with which she feared she could not cope. Only the confrontation with the high HbA1c value corresponding to an averaged blood glucose concentration of approximately 300 mg% convinced her to do what was indicated from the medical point of view.

This example demonstrates the importance of HbA1c determinations in uncovering manipulated results of self-monitoring.

A 23-year-old student of medicine became affected by insulin-dependent diabetes. In the last two years his HbA1c values were in the range of 6.2 and 6.5%. Occasional hypoglycemic episodes were mentioned but taken lightly. Upon inquiry the parents reported the repeated occurrence of severe hypoglycemic episodes in part with loss of conciousness. After such episodes a retrograde amnesia existed for the event so that the patient underestimated the risk of such occurrences. The patient resisted accepting a lower insulin dose and somewhat increased blood glucose concentrations for quite some time because his unrealistic goal was attainment of euglycemia because of fear of the late complications of this disease.

This example shows that glycosylated hemoglobin values in or close to the normal range may be associated with frequent and even severe hypoglycemic episodes. In these cases the insulin dose and schedule of injection have to be reevaluated.

As a diagnostic tool, estimation of glycosylated hemoglobin is not as discriminating as the standard oral glucose tolerance test or even the fasting blood glucose concentration[58]. Nevertheless, its convenience and lack of external variables may have some merit as a screening test.

4.3 Serum Protein Glycosylation

Estimation of total glycosylated plasma proteins or of their constituents such as albumin may provide information complementary to that obtained by HbA1 or HbA1c measurements. The short half-life of these proteins (e.g., 20 days for albumin) allows assessment of glycemic control over 1–2 weeks. This may be of value in pregnancy or when therapy is changed. Also there may be methodological advantages in it as total serum protein glycosylation can be measured with the rapid fructosamine test[39]. In the regular monitoring of glycemic control, measurement of serum protein glycosylation is of doubtful use because of the frequent sampling required for complete documentation.

5 Lipid Metabolism

Diabetes is often associated with hyperlipemia, a condition which appears to contribute to the accelerated atherosclerosis seen in diabetic patients. Hyperlipemia in diabetes most often develops as a consequence of insufficient metabolic control. This type of secondary hyperlipemia corresponds to the Type IV according to the nomenclature of Fredrickson[59].

While plasma triglyceride levels can be normalized with satisfactory blood glucose control[60], reduction of total plasma and LDL cholesterol and increases in HDL cholesterol may require near normalization of blood glucose control, as can now be achieved with the insulin pump[61].

Apart from occasional determination of ketone bodies, the measurement of lipid metabolism is not vital for the daily surveillance of the diabetic condition. As a consequence no methodological developments have occurred and methodology and criteria for lipid determinations are the same as in the diagnosis of disorders of lipid metabolism.

5.1 Ketone Bodies

Ketogenesis is accelerated under conditions where lipolysis and therefore nonesterified fatty acids in plasma are increased. This condition arises in marked insulin deficiency or when carbohydrate supply is lacking. The ketone bodies produced by the liver are acetoacetate, its product of reduction, 3-hydroxybutyrate, and that of spontaneous decarboxylation, acetone. Their proportions are variable, the concentration of the major ketone body 3-hydroxybutyrate varying between two and four times that of acetoacetate while acetone comprises only a few percent. In nondiabetics up to 100 μmol/l (1 mg%) ketone bodies are found in blood and only traces are excreted in the urine. During fasting and starvation plasma levels may increase 5–30 fold, while during ketoacidosis the increase is more than 100-fold in plasma and great amounts of ketone bodies are excreted in the urine.

Ketone bodies are usually measured semiquantitatively with the nitroprusside reaction[62)]. Nitroprusside reacts in alkaline solution with acetoacetate, and to a lesser extent also with acetone, to form a purple color. Both the Ketostix R paper strip and the Acetest R tablet make use of this reaction. The sensitivities of the tablet and strip are essentially identical and lie at approximately 5 mg/100 ml. Both forms of the test have color charts with three colors representing different ranges of positivity (+: 6–20 mg%/++: 20–50 mg%/+++: >50 mg% acetoacetate)[63, 64)]. In general, Ketostix read 100% positive in urine when the ketone body concentration in blood reaches 1 mmol/l. The accuracy of this semiquantitative methodology can be increased by preparing dilution series of plasma. In severe ketoacidosis the 1:4 dilution still reads ++ or +++.

Underestimation of ketonemia can occur with the nitroprusside methodology when the ratio acetoacetate: 3-hydroxybutyrate is shifted towards 3-hydroxybutyrate as is the case in combined lactic- and ketoacidosis[65)]. Because 3-hydroxybutyrate is the main ketone body and is the first to increase, the determination of 3-hydroxybutyrate is advantageous from the point of view of evaluating ketone body status. Its simple semiquantitative determination is possible with a paper test strip. This test strip contains the enzyme 3-hydroxybutyrate dehydrogenase and an indicator dye[66)].

There are some important limitations to using the determination of ketone bodies in urine as a measure of plasma ketone bodies. Increased urinary ketone body concentrations may occur without a concomitant increase of the plasma concentration[63, 67)]. On the other hand, in renal insufficiency, ketone body excretion may be absent despite elevated plasma ketone body concentrations.

The ketone body concentrations in blood and tear fluid are closely correlated, the concentration in the tear fluid being approximately half that in blood. Determination of ketone bodies in tear fluid is a valuable procedure in emergency situations[68)]. To this end, the Ketostix strip is carefully placed in the lower conjunctival sack until the entire reaction field is wetted. The strip is read 30 s later. A medium to strong reading (violet color) is indicative of a strong ketonemia generally associated with acidosis.

Clinically, urinary ketone body determinations provide a useful supplement to self-monitoring of daily urine glucose and allow recognition of a progressive metabolic decompensation. For the physician the ketone body determination in plasma or tear fluid is helpful in the bedside diagnosis of ketoacidosis. Although ketonemia is a useful parameter for the diagnosis of ketoacidosis it is inadequate for its evaluation because during the treatment of ketoacidosis there is usually an initial increase of the acetoacetate which may persist at a high level for several hours despite improvement of the patient and falling 3-hydroxybutyrate and glucose concentrations[69].

6 Residual Pancreatic Function: C-Peptide

The difficult decision whether recompensation of the metabolic state in type II diabetic patients with severely increased blood glucose concentrations (so called oral drug failure) can be achieved with the primary therapeutic measures (diet, weight reduction in obese patients, and oral antidiabetic drugs) or whether insulin therapy is indicated is facilitated by the evaluation of the residual pancreatic function[70].

Insulin secretion can be assessed by a C-peptide determination in plasma before and after beta cell stimulation with glucagon. Plasma C-peptide concentrations are determined by radioimmunoassay[71]. The glucagon-C-peptide test is performed after overnight fasting. The C-peptide concentrations are measured before and 6 minutes after intravenous injection of 1 mg glucagon.

Normal C-peptide values for nondiabetic, nonobese people are: 350 pmol/l (range 250–630 pmol/l) and increase to 1200 pmol/l (range 860–1880 pmol/l) after stimulation. It has been proposed that metabolic compensation without insulin requires basal C-peptide values >200 pmol/l and stimulated values in the range >600 pmol/l. The sensitivity of the glucagon-C-peptide test was 0.87, and its specificity 0.71[72,73]. These figures depend on the definition of oral antidiabetic drug failure. Opinion varies on this issue. Consensus exists that fasting whole blood glucose concentrations >11 mmol/l indicate sulfonylurea failure and that fasting

Table 7. Interpretation of the glucagon-C-peptide test: Values required for successful oral antidiabetic drug therapy

body weight (Broca)[a]	whole blood glucose, fasting (mmol/l)	necessary C-peptide concentration (pmol/l)	
		fasting/basal	after stimulation (1 mg glucagon)
<1.0	5– 7	250–400	> 650
	7– 9	400–550	> 800
	9–11	550–700	> 800
	>11	>700	> 800
1.0–1.25	5– 7	350–500	> 700
	7– 9	500–600	> 700
	9–11	600–800	> 700
	>11	>800	>1000
>1.25	5– 7	450–600	>1000
	7– 9	600–700	>1000
	9–11	700–900	>1000
	>11	>900	>1000

[a] Broca index $= \dfrac{\text{body weight (kg)}}{\text{body height (cm)} - 100}$

whole blood glucose <7.8 mmol/l are characteristic of responders. In the range 7.8–11 the decision is often difficult, especially in older people. Other criteria like impaired psychological and physical activity in combination with a depressive mood can be indicative of poor metabolic compensation necessitating insulin therapy.

A reevaluation in our patients indicates that discrimination of the glucagon-C-peptide test is best if blood glucose and body weight are also considered[74]. We now use the limits shown in Table 7 to evaluate results of the glucagon-C-peptide test. With these criteria, test sensitivity and specificity with regard to oral antidiabetic drug responsiveness are 0.9 for a 3-year observation period.

7 Microalbuminuria

The detection of microalbuminuria, that is the urinary albumin excretion below the level of "Albustix" detection, is gaining popularity because it allows detection of early renal damage at a stage when this complication may still be amenable to correction by careful maintenance of near-euglycemia and normotension[75, 76].

Diabetic patients with urine positive to the Albustix test generally have albumin excretion rates in excess of 200–250 mg/24 h (~150 μg/min). In healthy individuals, the albumin excretion rate varies between 2.5 and 26 mg/24 h (2–18 μg/min), the nocturnal rate being 40–50% lower than that during the day[77]. *Microalbuminuria* is defined as conditions with albumin excretion rates in the range between the physiological and clearly pathological albumin excretion, i.e., from 26–250 mg/24 h (18–174 μg/min). Clinical data show that overnight albumin excretion rates above 35 μg/min are associated with a 20-fold increased risk for nephropathy[75, 78].

Such small albumin concentrations in urine can be quantitatively estimated by radioimmunoassay or ELISA[79–82]. For screening purposes, a semiquantitative agglutination test which is sensitive, simple, and quick to perform may be used[83]. It gives results in a concentration range between 2.5 and 17 mg/dl, and reads as negative concentrations in the normal range and those above 17 mg/dl. The latter, however, are read as positive by the Albustix test.

Tools are thus available both for mass identification of patients with microalbuminuria, and for precise quantitation of their urinary albumin excretion.

8 Evaluation of Metabolic Control

Results of self-monitoring of metabolic control improve the physician's view of the course of the disease and facilitate decisions regarding therapeutic measures. However, the full profit of self-monitoring shows only if it enables the patient to adjust the therapy to the actual situation. This requires understanding of the basic metabolic relationships. Visualization of test results in an organized manner is an important step in this direction. The format of the protocol used in our labile

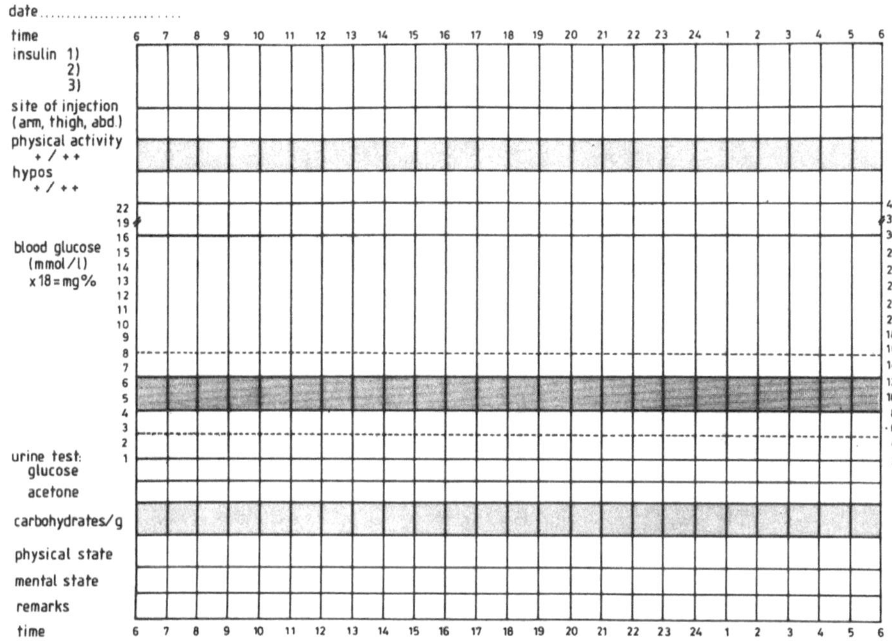

Fig. 8. Example of data-sheet format for recording results of self-monitoring

diabetic patients is shown in Fig. 8. In this form the hypo- and hyperglycemic range is clearly marked and values outside this range require reconsideration of therapy.

Recently a diabetes management program for a personal computer has become available which carries out data reduction and even adjustment of insulin therapy. With the increasing number of private personal computers, this tool may have great potential for facilitating the management of the diabetic condition by both patient and physician[84].

8.1 Guidelines for Surveillance and Evaluation of Therapy

The criteria for the evaluation of metabolic control in diabetic patients are summarized in Table 8 and the recommended test frequencies in Table 9.

8.1.1 Assessment of the Risk for Hypoglycemia in Type I Diabetics

Two recent studies have shown that the occurence of asymptomatic nocturnal hypoglycemia is unexpectedly high as one third of type I diabetic patients experiences such episodes most of which remain asymptomatic [4, 85]. The risk for nocturnal hypoglycemia can be estimated from the blood glucose concentration at bedtime. Normal blood glucose values at bedtime (below 6 mmol/l) were associated in 80% with nocturnal hypoglycemia. Regular determination of the blood glucose concen-

Table 8. Criteria for the evaluation of metabolic control

Assessment	Whole blood glucose (mmol/l)				Hemoglobin A1c/A1 (% tot. Hb)		Urinary glucose (g/24 h)	Keton-uria	Cholesterol/ triglycerides mmol/l (mg%)	Body weight (Broca)
	fasting preprandial and bedtime	after breakfast 1 h	2 h	after lunch or supper 3–5 h						
Poor	>7.8	>14	>10	>7.8	>7.5	>10	>15	–/+	<6.2 (<240) <2.3 (<200)	>1.1
Satisfactory	<7.8 >5.0[a]	<14	<10	<7.8 >5.0	<7.5	<10	<15	–	<5.7 (<220) <1.7 (<150)	1–1.1
Near normal[b]	4–5.5	5.5–7.8	5.5–6.7	4.4–5.5	<6.0	<8.3	neg.	–		0.8–1.0

[a] With *insulin therapy* and *sulfonylurea therapy* fasting and bedtime blood glucose concentrations below 5 mmol/l are associated with an increased risk of nocturnal hypoglycemia

[b] Near normal glycemia is desirable during pregnancy and in juvenile diabetics after onset of diabetes. In most of the patients on insulin therapy for more than 10 years and in brittle diabetes, achievement of near normal glycemia is not possible with conventional insulin therapy because of the risk of severe hypoglycemia

Table 9. Basic metabolic parameters in the routine management of diabetes

	Parameter				
	Urinary glucose		Blood glucose		Hemoglobin A1c
Therapy	When	How often	When	How often	How often
Diet	after breakfast	daily	fasting, 1-h postprandial	quarterly	half-yearly
Oral anti-diabetic drugs	after breakfast	daily	fasting, 3–5 h after lunch (1-h postprandial)	quarterly	half-yearly
Insulin (stable metabolic control)	before principal meals	daily	fasting, 1-h postprandial, before lunch and supper (at night 2 a.m.)	every 4–6 weeks	half-yearly
Insulin (labile metabolic control)	before principal meals and at bedtime	daily	fasting, before lunch, supper, and bedtime, 1-h postprandial, at night 2 a.m.	every 2–4 weeks[a]	at least every 3rd month

Ketone determinations only in special situations like impaired well-being and heavy glycosuria (>1%) over 24 h

[a] During intensified insulin treatment (multiple daily injections or pump therapy) daily measurements are necessary

tration at bedtime is therefore essential to recognize the risk for nocturnal hypoglycemia and for proper adjustment of therapeutic measures. As the lowest blood glucose concentrations occur between 02 and 04 a.m., periodical measurement of blood glucose values at this time reveals the individual risk for hypoglycemia.

9 Conclusions and Perspectives

During the last decade there has been a development of methodology which allows objective monitoring of metabolic control in diabetes. In many instances application of these methods results in the improvement of glycemic control to a near-normal range. However, in some cases the intrinsic limitation of the conventional insulin therapy does not allow patient and physician to achieve optimal compensation of metabolic control. It should also be considered that in the achievement of good glycemic control patient cooperation, motivation, and adherence to prescribed therapy determine the success of therapy. This important aspect has been documented by a controlled trial showing that intensive attention can improve glycaemic control without further advantage from home blood glucose monitoring[86].

The main limitation of existing methodology is that it does not allow continual monitoring of blood glucose concentrations. This would be essential to detect asymptomatic nocturnal hypoglycemia and to optimally adapt therapy to the needs of the body as would be realizable by a closed loop system (artificial pancreas). The realization of glucose sensors which allow continuous blood glucose monitoring will be a major step towards blood glucose normalization. It is encouraging that preliminary reports on the realization of such sensors have already appeared[87].

10 Acknowledgments

The authors thank Dr. T. Resink and N. Iberg for revising the manuscript. One of the authors (R.F.) was a recipient of a grant of the Johnson Foundation.

11 References

1. Constam, G. R.: Med. Welt *26,* 1802 (1975)
2. Engerman, R.: Diabetes *26,* 760 (1977)
3. Pirart, J.: Diabete Metab. *3,* 97–, 173–, 245– (1977) and Diabetes Care *1,* 168–, 252– (1978)
4. Vonesch, H. J. et al.: Schweiz. med. Wochenschr. *4,* 1674 (1984)
5. Berger, W.: Med. et Hyg. *40,* 1407 (1982)
6. Pileggi, V. J., Szustkiewicz, Ch. P.: Carbohydrates, in: Clinical Chemistry, Principles and Techniques (ed.) Henry, R. J. et al., p. 1265, New York, Harper & Row, 2nd edition, 1974
7. Sunderman, F. W. Jr., Sunderman, F. W.: Am. J. Clin. Path. *36,* 75 (1961)
8. Mager, M., Farese, G.: Am. J. Clin. Path. *44,* 104 (1965)
9. Dillon, R. S.: Diabetes *14,* 672 (1965)
10. Overfield, C. V.: Clin. Chim. Acta *39,* 35 (1972)
11. Ruiter, J., Weinberg, F., Morrison, A.: Clin. Chem. (NY) *9,* 356 (1963)
12. WHO Expert Committee on Diabetes Mellitus. Second Report. Technical Report Series 646. WHO Geneva (1980)
13. Studer, E., Cikes, M., Widmer, L. K.: Dtsch. Med. Wochenschr. *22,* 1179 (1969)
14. Suarez, L., Barrett-Connor, E.: Diabetologia *22,* 250 (1982)
15. National Diabetes Data Group: Diabetes *28,* 1039 (1979)
16. O'Sullivan, J. B., Mahan, C. M.: Diabetes *13,* 278 (1964)
17. Lundbaek, K.: Br. Med. J. 1507 (1962)
18. Reeves, M. et al.: Diabetes Care *4,* 404 (1981)
19. Frey, H. O. et al.: Münch. Med. Wochenschr. *46,* 1545 (1979)
20. Clements, R. S. et al.: Diabetes Care *4,* 392 (1981)
21. Schiffrin, A., Desrosier, M., Belmonte, M.: Diabetes Care *6,* 166 (1983)
22. Lombrail, P. et al.: Diabetes Care *7,* 318 (1984)
23. Schlichtkrull, J., Munck, O., Jersild, M.: Acta. Med. Scand. *177,* 95 (1965)
24. Service, F. J. et al.: Diabetes *19,* 644 (1970)
25. Hayford, J. T., Weydert, J. A., Thompson, R. G.: Diabetes Care *6,* 40 (1983)
26. Service, J. F. et al.: JAMA *222,* 294 (1972)
27. Klemmer, L., Thoma, J., Kutter, D.: Med. Lab. *2,* 39 (1982)
28. Banauch, D., Koller, P. U., Bablok, W.: Diabetes Care *6,* 213 (1983)
29. Feldman, J. M., Kelley, W. N., Lebovitz, H. E.: Diabetes *19,* 337 (1970)
30. Feldman, J. M., Lebovitz, H. E.: Diabetes *22,* 115 (1973)

31. Walford, S., Page, McB, Allison, S. P.: Diabetes Care *3,* 672 (1980)
32. Johansen, K., Svendsen, P. A., Lorup, B.: Diabetologia *26,* 180 (1984)
33. Fine, J.: Br. Med. J. *1,* 205 (1967)
34. Renschler, H.: (Habilitationsschrift)
35. Bunn, H. F., Gabbay, K. H., Gallop, P. M.: Science *200,* 21 (1978)
36. Flückiger, R.: Lab. Med. *8,* 120 (1984)
37. Schleicher, E., Wieland, O. H.: J. Clin. Chem. Clin. Biochem. *19,* 81 (1981)
38. Mallia, A. K. et al.: Anal. Lett. *14,* 649 (1981)
39. Baker, J. R. et al.: Br. Med. J. *290,* 352 (1985)
40. Flückiger, R., Gallop, P. M.: Methods Enzymol. *106,* 77 (1984)
41. Mayer, T. K., Freedman, Z. R.: Clin. Chim. Acta *127,* 147 (1983)
42. Javid, J. et al.: Br. J. Haematol. *38,* 329 (1978)
43. Gabbay, K. H. et al.: J. Clin. Endocrinol. Metab. *44,* 859 (1977)
44. Flückiger, R., Pasquel, M.: Schw. Med. Wochenschr., in press
45. Bisse, E., Berger, W., Flückiger, R.: Diabetes *31,* 630 (1982)
46. Goldstein, D. E. et al.: Diabetes *31* (Suppl. 3), 70 (1982)
47. Phelps, R. L. et al.: Am. J. Obstet. Gynecol. *147,* 651– (1983)
48. Popat, N. et al.: Lancet *ii,* 377 (1977)
49. Sosenko, J. M. et al.: Diabetes Care *3,* 590 (1980)
50. Flückiger, R. et al.: New Engl. J. Med. *304,* 823 (1981)
51. Hoberman, H. D., Chido, S. M.: Alcoholism *6,* 260 (1982)
52. Bridges, K. R. et al.: J. Clin. Invest. *56,* 201 (1975)
53. Panzer, S. et al.: Klin. Wochenschr. *61,* 839 (1983)
54. Falko, J. M., O'Dorisio, T. M., Cataland, S.: Arch. Intern. Med. *142,* 1370 (1982)
55. Svendsen, P. A. et al.: Diabetologia *23,* 403 (1982)
56. Nathan, D. M. et al.: New Engl. J. Med. *310,* 341 (1984)
57. Berger, W., Flückiger, R.: Schweiz. Aerztezeitung *34,* 1620 (1984)
58. Kesson et al.: Diabetes Care *5,* 395 (1982)
59. Fredrickson, D. S., Levy, R. I., Lees, R. S.: New Engl. J. Med. *276,* 34 (1967)
60. Lopes-Virella, M. F. et al.: Diabetes *32,* 20 (1983)
61. Pietri, A., Dunn, L. F., Raskin, Ph.: Diabetes *29,* 1001 (1980)
62. Free, A. H., Free, H. M.: Am. J. Clin. Pathol. *30,* 7 (1958)
63. Watkins, P. J., Fitzgerald, M. G.: Diabetes *17,* 398 (1968)
64. James, R. C., Chase, G. R.: Diabetes *23,* 474 (1974)
65. Marliss, E. B. et al.: New Engl. J. Med. *283,* 978 (1970)
66. Harano, Y. et al.: Diabetes Care *7,* 481 (1984)
67. Coetzee, E. J., Jackson, W. P. U., Berman, P. A.: Diabetes *29,* 177 (1980)
68. Berger, W., Goeschke, H.: Med. et Hyg. *31,* 1 (1973)
69. Stephens, J. M., Sulway, M. J., Watkins, P. J.: Diabetes *20,* 485 (1971)
70. Peacock, I., Tattersall, R. B.: Br. Med. J. *288,* 1956 (1984)
71. Heding, L. G.: Diabetologia *11,* 541 (1975)
72. Madsbad, S. et al.: Acta Med. Scand. *210,* 153 (1981)
73. Poulsen, S. et al.: Horm. Metabol. Res. *17,* 39 (1985)
74. Pasquel, M. et al.: Schweiz. Med. Wochenschr., in press
75. Viberti, G. C. et al.: Lancet *i,* 1430 (1982)
76. Morgensen, C. E.: New Engl. J. Med. *310,* 356 (1984)
77. Morgensen, C. E.: Scand. J. Clin. Lab. Invest. *28,* 101 (1971)
78. Parving, H.-H. et al.: Acta Endocrinol. (Copenhagen) *100,* 550 (1982)
79. Keen, H., Chlouverakis, C.: Lancet *ii,* 913 (1963)
80. Miles, D. W., Morgensen, C. E., Gundersen, H. J. G.: Scand. J. Clin. Lab. Invest. *26,* 5 (1970)
81. Woo, J. et al.: Clin. Chem. (NY) *24,* 1464 (1978)
82. Fielding, B. A., Price, D. A., Houlton, C. A.: Clin. Chem. (NY) *29,* 355 (1978)
83. Viberti, G. C., Vergani, D.: Diabetes *31,* 973 (1982)
84. Rodbard, D., Pernick, N., Jaffe, M. L.: Diabetes Care *7,* 401 (1984)
85. Pramming S. et al.: Br. Med. J. *291,* 376 (1985)
86. Worth, R. et al.: Br. Med. J. *285,* 1233 (1982)
87. Shichiri, M.: Diabetes *33,* 1200 (1984)

Recent Chemical Developments in the Field of Beta Adrenoceptor Blocking Drugs*

Herbert G. Köppe

Boehringer Ingelheim KG, Department of Medicinal Chemistry
D-6507 Ingelheim am Rhein, West Germany

β-adrenoceptor blocking drugs have been recognized agents in the therapy of heart diseases and vascular hypertension for years. Since their discovery, they have had an unusual career in the heart and circulation sector. Initially, their administration was restricted to the treatment of angina pectoris. After an antihypertensive effect had been observed in patients with angina, they were also used in hypertension.

Variations of the chemical structure, mainly in the class of 1-phenoxy-3-alkylamino-2-propanols, have yielded agents with specific action profiles. β-adrenoceptor blocking drugs with intrinsic sympathomimetic activity (ISA) as well as substances with a relatively selective action on β_1-receptors were discovered.

In recent years, the focus of research shifted to stronger and longer acting antihypertensives. Current synthesis is aimed at producing β-adrenoceptor blocking drugs which have a vasodilatory or diuretic effect in addition.

* Presented in part at the annual meeting of the Gesellschaft Deutscher Chemiker, Medicinal Chemistry Section, Freiburg, September 5, 1984

Progress in Clinical Biochemistry
and Medicine, Vol. 3

1 Introduction

In the last twenty years, the importance of β-adrenoceptor blocking drugs in the treatment of cardiovascular diseases has been increasing steadily. Their development has been among the most important advances in pharmacotherapy in the last two decades[1–4]. In the treatment of both coronary heart disease[5] as well as of certain tachycardiac forms of arrhythmias[6], they are regarded as an established component and as a major improvement. β-adrenoceptor blocking drugs have also been used in the treatment of hypertension[7, 8] after overcoming the reservations which prevailed for a time, and today, they are now standard therapy[9] in the treatment of hypertension. In addition, further possible applications in the CNS region and in glaucoma have appeared[10].

Increased and differentiated therapeutic use and molecular modifications of β-blockers have led to the theory of the existence of β_1- and β_2-adrenoceptors[11, 12]. This will be dealt with below (pp 37, 38).

Pharmacological results support the assumption that β_1-adrenoceptors predominate in various organs, whereas only small β_2-adrenoceptor populations are present.

Although it was possible to synthesize substances which selectively occupy β_1-adrenoceptors, the hope of achieving a pure β_1 effect in these organs with such substances has not been completely fulfilled. The term "relative β_1-receptor selectivity" was coined. According to present-day knowledge, all the therapeutic effects mentioned above are mediated by the blockade of β_1-adrenoceptors. Consequently, administration of a selective β_1-blocker should be given preference.

A further feature distinguishing the β-blockers is associated with the concept of intrinsic sympathomimetic activity (ISA). A β-blocker with ISA is an agent with a dualistic mode of action. Besides a mainly antagonistic effect, it is also able to exert an agonistic effect. β-blockers with ISA are indicated for the therapy of cardiovascular diseases, in which a marked lowering of heart rate is undesirable. However, the clinical significance of this type of β-blocker has still not been completely clarified[13].

A nonspecific, quinidine-type membrane effect has been reported as a special feature of certain β-blockers. This is a blocking effect on cellular sodium influx. It is connected with a local anesthetic effect. Today, however, it is virtually certain that this effect is insignificant in therapy with β-blockers since contrary to earlier views this effect either occurs only rarely[14, 15] or does not lead to clinical effects in conventional therapeutic doses[16].

A large number of β-blockers has been developed by the pharmaceutical industry in recent years and made available for treating the syndromes mentioned. However, the exploration of these substance classes with uniform or similar basic chemical structure cannot yet be regarded as completed.

The following compilation, preceded by a historical review, provides a survey of recent research activities throughout the world. From the vast number of variations in chemical structure which have been documented mainly in patents and patent applications, a strictly limited selection of studies which appear to be important has been made. Marketed β-blockers of the first generation are not discussed apart from those which have served as structural "leads" for molecular variations. In this connection, the reader is referred to comprehensive chemical[12, 18–20] and pharmacological/medical[2, 3, 7, 10, 14, 21, 22] reviews of this research field and the indications.

2 Historical Review

In his classical study in 1948[23], R. P. Ahlquist developed the theory that the adrenergic transmitters adrenaline and noradrenaline exert their effects via two different receptor types which he termed α- and β-receptors. His work constitutes the basis for the understanding of β-receptor blockade.

In the 1950s, during the search for new bronchodilators, studies on phenethanolamine structures related to isoproterenol were carried out in the Lilly research laboratories (Shanks)[24]. Among other substances, 1-(3,4-dichlorophenyl)-2-isopropylamino-ethanol (Mills)[25] was synthesized, which has become known under the designation dichloroisoproterenol or dichloroisoprenaline = DCI. It differs from isoproterenol in that its two hydroxyl groups in the benzene ring are replaced by two chlorine atoms.

The pharmacological studies on the mode of action of DCI surprisingly showed that it elicits an effect antagonistic to that of isoproterenol (Powell and Slater, Moran and Perkins)[26, 27]. However, since continuing agonistic effects were observed besides the isoproterenol antagonism, research with this first β-blocker of the phenethanolamine group was discontinued.

A little later (in June 1960), novel 1-alkylamino-3-phenoxy-2-propanols were synthesized at Boehringer Ingelheim (Köppe)[28]. The objective of these syntheses was to obtain anorectic substances without a central stimulant side effect such as had been observed after administration of the anorectic phenmetrazine. Phenmetrazine can be regarded as a cyclized phenethanolamine derivative. Among other drugs, the phenethanolamine derivatives developed at Boehringer Ingelheim include the antiasthmatic drug isoproterenol (Aludrin®) (Scheuing, Thomae)[29]. Isoproterenol was the first substance with which it was possible to bring about a stimulation of the β-adrenergic receptors. So, for the first time, it was possible to distinguish between α- and β-adrenergic receptors. The common structural features and structural differences between the above-mentioned substances are shown in Fig. 1.

During the synthesis and pharmacological testing of the 1-alkylamino-3-phenoxy-2-propanols for anorectic activity the studies on DCI became known. Although, as phenol ethers (phen–O–CH_2–CH<), these substances substantially differ chemically from the phen–CH<-linked phenethanolamines (e.g. DCI), it was concluded on the basis of comparable structural features that such compounds might also elicit effects antagonistic to isoproterenol (Köppe, Engelhardt)[30]. Pertinent tests were planned and carried out at the beginning of 1961 with the 1-ethylamino-3-(2,4,5-trichlorophenoxy)-2-propanol (Kö 446) (Engelhardt). Decisive for the choice of this substance was 3,4-dichloro substitution also shown by DCI. The structure of this compound is compared to that of DCI in Fig. 1. In contrast to most substances in which isoproterenol antagonism was later found, the amino group is occupied by the unbranched ethyl group instead of a branched alkyl residue. Precisely for this reason and because of substantial structural differences due to the –O–CH_2-bridge of the substance Kö 446 and a further chlorine substitution, the test results were surprising. It was observed in guinea-pigs (Konzett-Rössler) that the bronchospasmolytic action of isoproterenol and adrenaline on acetylcholine spasm is inhibited. A negative inotropic and a negative chronotropic intrinsic effect were detected in the iso-

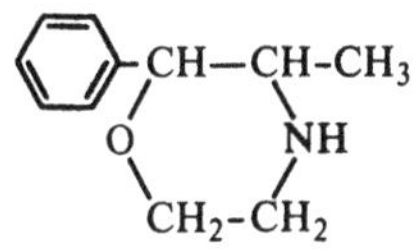

Phenmetrazine
Boehringer Ingelheim (1950) 1)

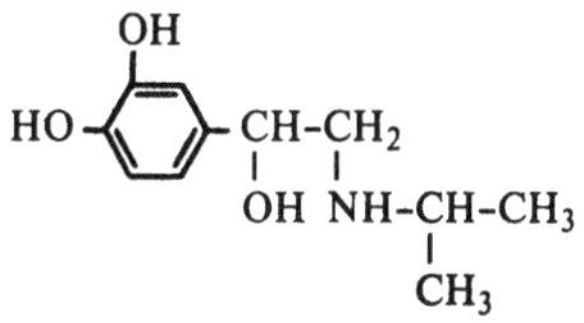

Isoproterenol
Isoprenaline
Boehringer Ingelheim (1939) 2)

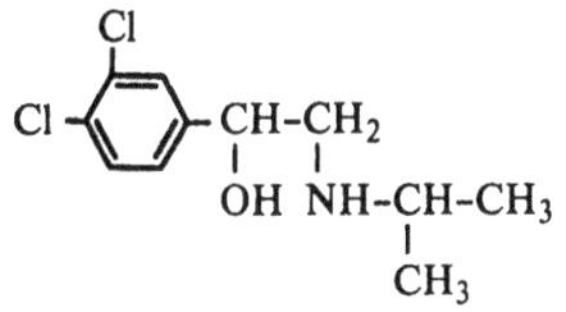

Dichloroisoproterenol
Dichloroisoprenaline
(DCI) Lilly (1958) 3)

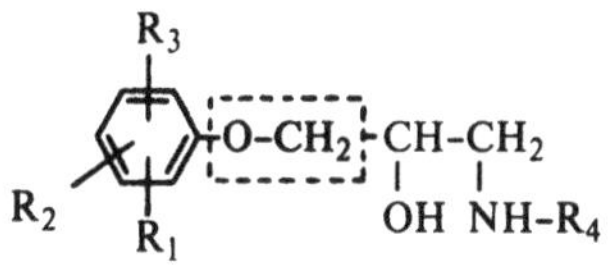

1-Alkylamino-3-phenoxy-2-propanols 4)
Kö 446: $R_1, R_2, R_3 = 2,4,5\text{-}Cl_3$;
$R_4 = C_2H_5$
Boehringer Ingelheim (1960)

Fig. 1 Structural relationships between anorectics, antiasthmatic and beta adrenoceptor blocking drugs

lated guinea-pig atrium. In trials repeated several times, antagonistic activity against the cardiac effect of isoprenaline (heart rate, amplitude) was observed. In May 1961, A. Engelhardt[31)] summarized the surprising and highly significant results of this test as follows:

"Kö 446 is a highly active β-adrenolytic. Its efficacy is between that of dichloroisoprenaline and dichloroadrenaline, but it differs from both these substances in that there is no intrinsic sympathomimetic activity. In a therapeutic trial with this substance, it would be possible to block a part of the sympathetic nervous system (the β-adrenoceptors) which it has so far not been possible to influence in terms of inhibition. Thus, further pharmacological investigation and possibly clinical testing appear to be called for."

Consequently, the substance Kö 446 was recognized and characterized as a β-adrenoceptor blocking drug. At the same time, it is the first β-blocker in the 1-alkylamino-3-phenoxy-2-propanol series. Moreover, the further course of development of this group of substances, which unexpectedly was to attain such eminence in the therapy of cardiovascular diseases, is indicated in the report on the pharmacological results.

These highly interesting results spurred on chemical and pharmacological studies. All 1-alkylamino-3-phenoxy-2-propanols synthesized since June 1960[32)] were immediately tested for isoproterenol antagonism. This activity was especially demonstrated by compounds in which the amino group shows a branched alkyl residue. Up to April 1962, the substances presented in Fig. 2 had revealed a pronounced β-blocking effect.

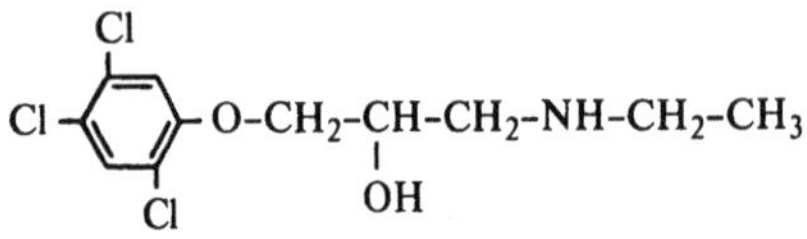

Kö 446 4)

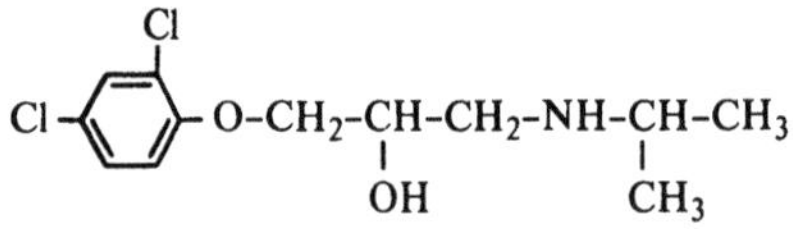

Kö 581 5)

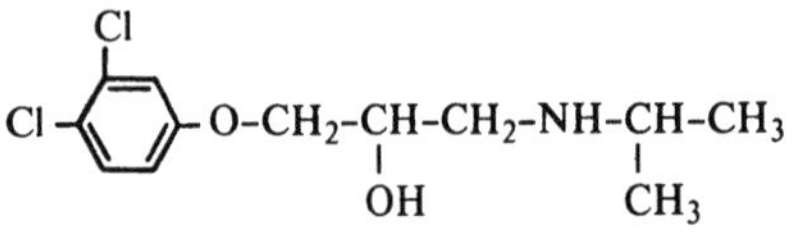

Kö 673 6)

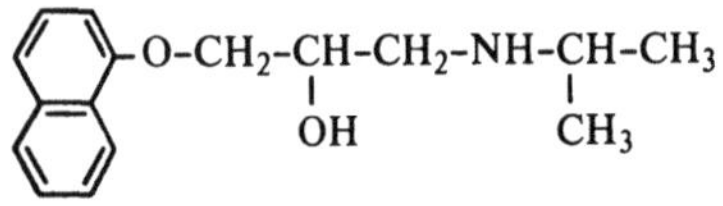

Lg 32 7)

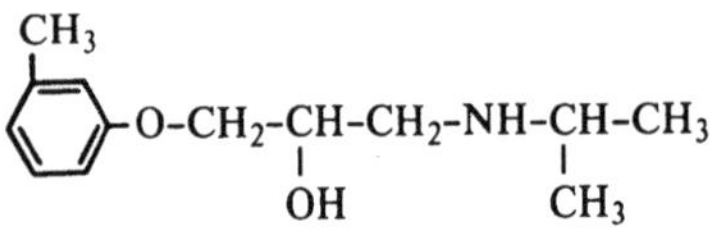

Kö 592 8)

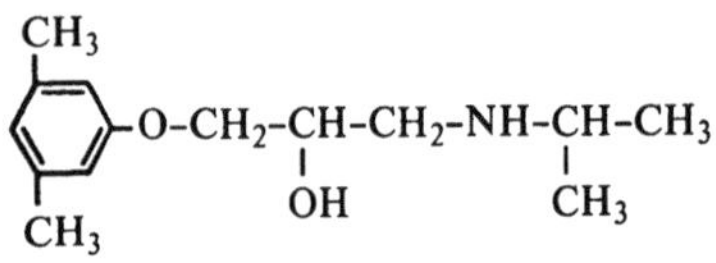

Kö 707 8a)

Fig. 2. Synthesis and detection of beta adrenolytic effective 1-alkylamino-3-phenoxy-2-propanols (1960–1962, Boehringer Ingelheim)

In a narrowed-down selection, the compounds Lg 32, which later became known as propranolol, and Kö 592 were favored. Since Kö 592 showed the least cardiodepressive action[33–36], this compound was developed under the generic name toliprolol and introduced in 1970 as Doberol®.

The historical development of β-blockers at research centers in the universities and industry is described in the excellent review by R. G. Shanks[1, 28].

3 β-Adrenoceptor Blocking Drugs as Antianginal Agents

Use of β-blockers in the pharmacotherapy of coronary heart disease is mainly aimed at reducing the oxygen requirement of the heart and balancing an inadequate relationship between oxygen requirement and supply.

3.1 β-Adrenoceptor Blocking Drugs with Intrinsic Sympathomimetic Activity (ISA)

Among the β-blockers displacing catecholamines from the β-adrenoceptor by competitive antagonism, we can distinguish between those β-adrenoceptor blocking drugs with and without intrinsic sympathomimetic activity (ISA), those with and without cardioselectivity (also termed β_1-selectivity) and those with and without nonspecific membrane action.

Figure 3 shows a comparison of β-blockers with and without intrinsic sympathomimetic activity.

It is considered that β-adrenoceptor blocking drugs without ISA should in general be used for the therapy of angina pectoris in order to economize the cardiac work. This is achieved inter alia by the bradycardiac effect of the β-adrenolytics. β-blockers with ISA appear to be appropriate in patients with a tendency to bradyarrhythmias. Despite numerous publications, the place of β-adrenoceptor blocking agents with ISA cannot yet be established. It should be noted that structure-activity relationships cannot be inferred for such preparations exhibiting a comparatively mild ISA. For this reason, little specific research work has been done in this field in recent years.

There has subsequently been an increase in chemical investigation of β-adrenolytics which exert a very pronounced ISA. The aim here is to obtain agents in which the ISA predominates, so that they can be used as positively inotropic substances in the treatment of myocardial insufficiency. Although partial β-agonists[37)], they cannot be used for the indications of β-antagonists. They are discussed in this context on

beta adrenoceptor blocking drugs	
without ISA	with ISA
Propranolol Toliprolol Atenolol Metoprolol Nadolol Timolol	Acebutolol Bunitrolol Alprenolol Carteolol Oxprenolol Pindolol

Fig. 3. Beta adrenoceptor blocking drugs and ISA

the basis of their chemical origin and the same basic molecular structure. Compounds of this type were already synthesized in 1962, and their inotropic effect was described (Köppe, Engelhardt)[38, 39]. As essential substituents, they show one or several hydroxy or methoxy groups in the phenyl ring of the β-blocker structure. These are responsible for the salient intrinsic sympathomimetic activity.

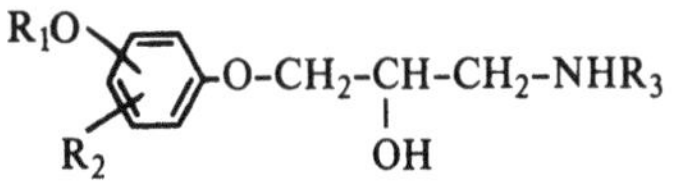

9)

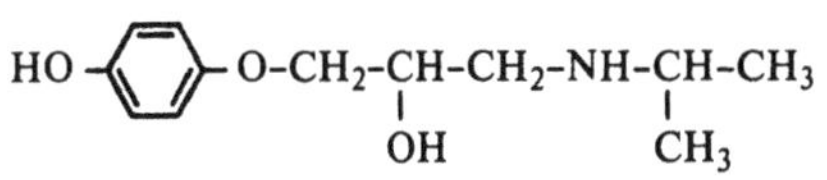

Kö 727
Boehringer Ingelheim
(1962)

10)

Fig. 4. Beta adrenoceptor blocking drugs with strong ISA

The S-enantiomer of the substance *(10)* (Kö 727) has been clinically developed under the generic name prenalterol[40–42].

Starting from the structure of prenalterol, an attempt was made to further enhance ISA by molecular variations and, on the other hand, to reduce the disadvantages of the available substances, e.g. the alteration of the β-receptor populations. Figure 5 shows a selection of new types of compounds and substances under development in which an increase in the intrinsic sympathomimetic properties could be achieved by modifying the alkylamino group of prenalterol.

In compounds *(11)* to *(15)*, the prenalterol skeleton has been preserved almost unchanged. By linking the alkylamino group of prenalterol to a phenyl residue in *(11)* with a hydroxy and a carbonamide group, hydrophilicity and inotropic properties are increased[43–45]. The situation is quite similar in compound *(12)*[46]. Analogously, the compounds *(13)* and *(15)*[50] which are substituted by a heterocycle at the terminal carbonamide group resemble each other. The orally active xamoterol is under clinical development[47]. Equally, this β_1-receptor-selective substance is a partial antagonist[48, 49]. In class *(14)* compounds, the hydroxylated phenoxyphenyl group contributes to the partial antagonistic action[51]. The class *(16)* benzofuran derivatives can be regarded as cyclized variants of the β-blocker molecule. The inotropic properties resemble those of prenalterol[52]. In substance *(11)*[44], β_1-selectivity was also observed besides the inotropic effect. This can be attributed to the typical substitution of the amino group.

S 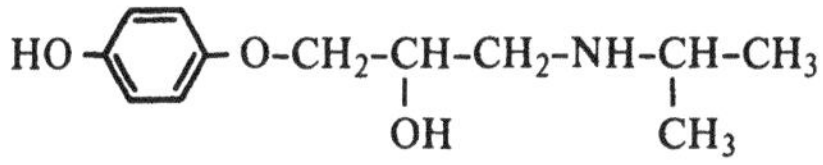

Prenalterol
Boehringer Ingelheim
Hässle
Ciba-Geigy
10)

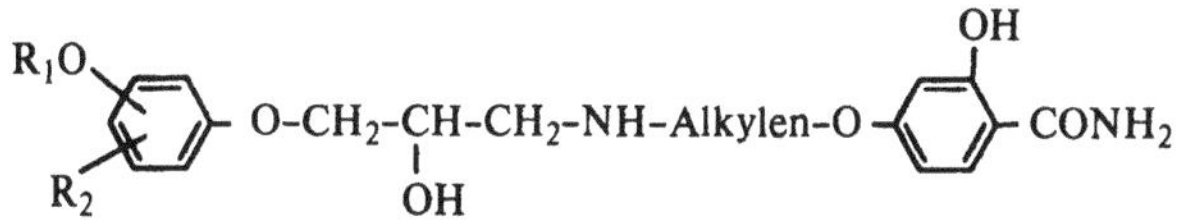

Ciba-Geigy 11)

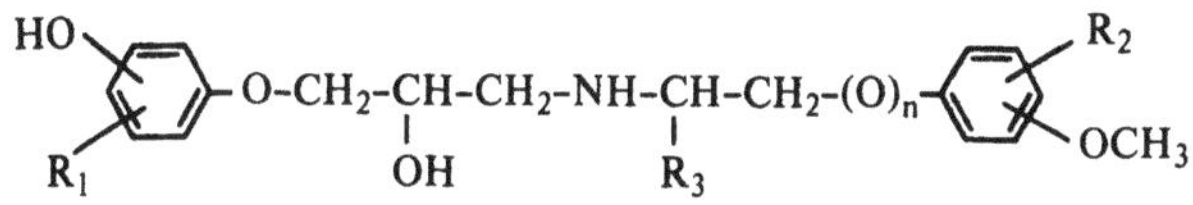

Hässle 12)

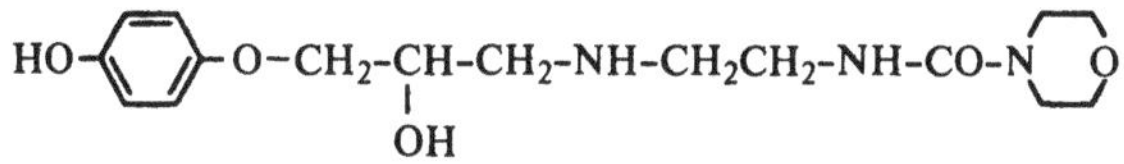

Xamoterol
(ICI 118587)
ICI
13)

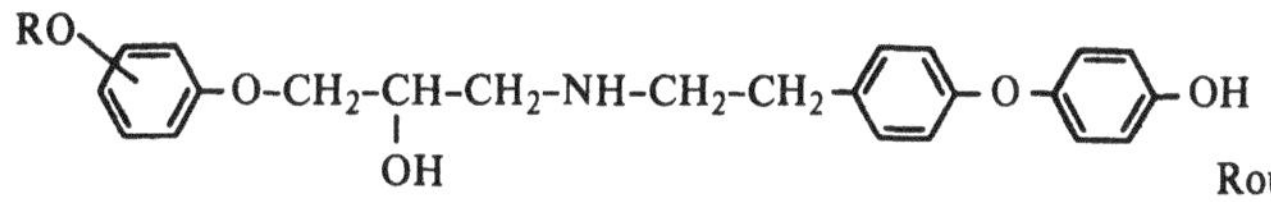

Roussel-Uclaf 14)

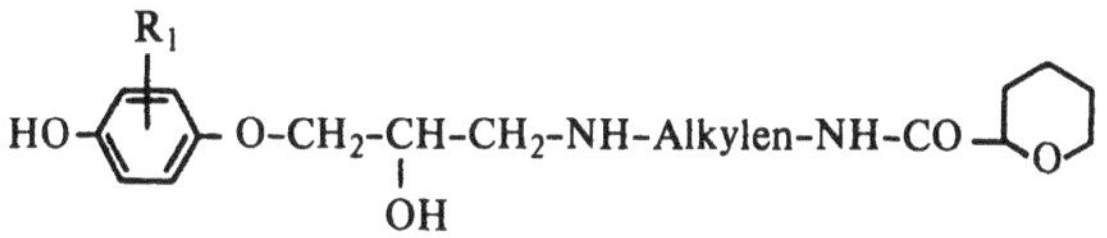

Louis 15)

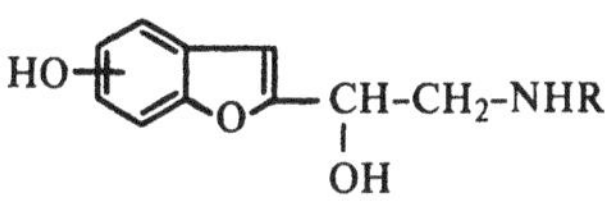

Hässle 16)

Fig. 5. New beta adrenoceptor blocking drugs with strong ISA

3.2 *β-Adrenoceptor Blocking Drugs and β_1-Adrenoceptor Selectivity*

In the early years of development of β-adrenoceptor blocking drugs, the term cardioselectivity did not yet exist. As we know today, the agents available at that time did not have the prerequisites of chemical structure in order to be able to elicit selective blockade of β-receptors in animal tests. Only after interest in the new agents had grown significantly and the skeleton of the 1-alkylamino-3-phenoxy-2-propanols had been altered even more extensively in the pharmaceutical research laboratories did substances become available which were predestined to exert their effect in a specific way. Since special attention had been paid to the development of β-adrenoceptor blocking drugs in England, it was only a question of time before the selectivity of agents with appropriate structures would be detected. The first cardioselective substance was described in 1967/68 (Shanks)[53] and later became known under the name of practolol. At the same time, the subdivision of the β-receptors into β_1- and β_2-receptors was suggested by Lands[54]. According to him, the β_1-receptors are located mainly in the heart, whereas the β_2-receptors are to be found in the smooth musculature, e.g. of bronchi, blood vessels and uterus. Although the new selective β_1-receptor blockers were supposed to block the β_1-receptors, it was recognized that an absolute cardioselectivity can not be expected because of the small number of β_2-receptors in the same area[55, 56].

In low doses, β_1-selective blockers occupy the β_1-receptors selectively because of their increased affinity for them. However, already in therapeutic doses this selectivity is often lost, at least to some extent, since β_2-receptors are also blocked at the same time. The term of "relative cardioselectivity" or "relative β_1-receptor selectivity" has thus come into use[57]. It is considered today that agents with β_1-selectivity should be given preference in certain indications, e.g. in asthma and diabetes[3, 58, 59]. Accordingly, it was the obvious next step for pharmaceutical research to concentrate its efforts in this direction in order to obtain further improved agents. These studies yielded β_1-selective adrenoceptor blockers which have served as guides for new syntheses in recent years. The chemical structure of the early developments can be seen in Fig. 6.

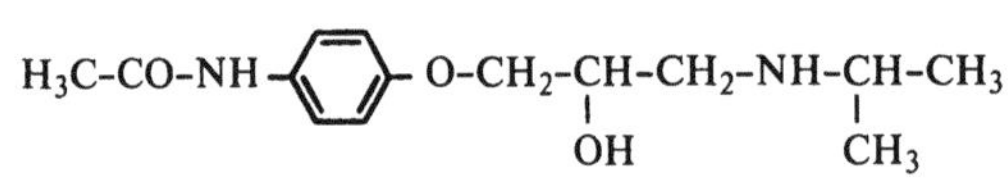

Practolol
ICI
17)

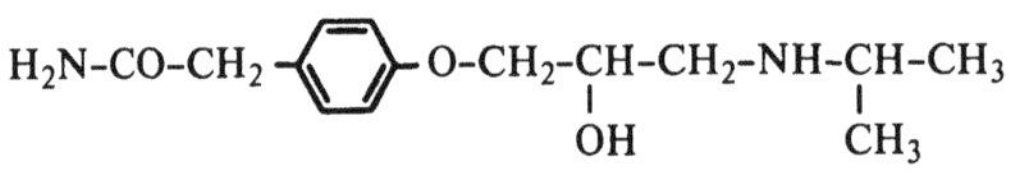

Atenolol
ICI
18)

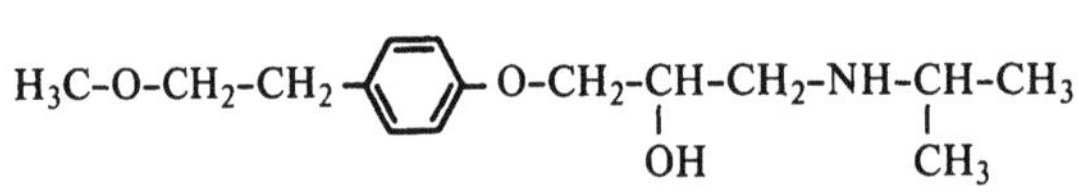

Metoprolol
Ciba-Geigy
19)

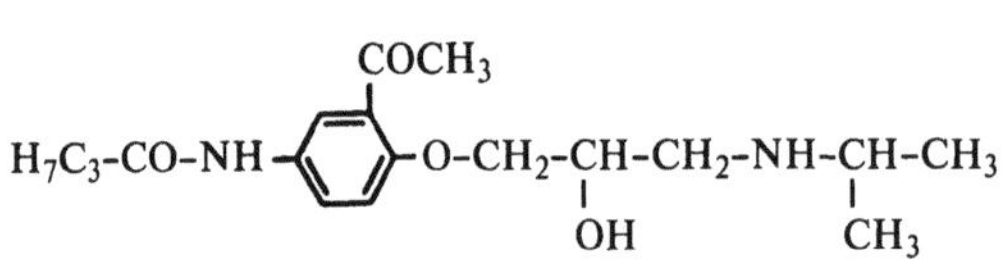

Acebutolol
May & Baker
20)

Fig. 6. Selective β_1-adrenoceptor blocking drugs

Comparison of the structures of these four selective β_1-receptor blockers makes it quite clear that their aliphatic side chain is completely identical. In connection with the phenoxy residue, this branched alkylamino isopropanol chain is essential for isoproterenol antagonism. Apart from very few β-blockers with tertiary linkage of the amino group, all β-blockers have a secondary alkylamino group. The substituents of the phenyl ring may be chemically different but they are always in the 4-position in relation to the ether oxygen in selective β-blockers[12, 60–62]. Hydrophilic or lipophilic and space-occupying properties of the substituents are of considerable importance and determine the degree of selectivity. A carbonamide[64–66] or ether oxygen function[67–69] is important for selectivity. Acebutolol[70] shows that a second substituent can be advantageous[71, 72]. A detailed description of the structure-activity relationships in β-blockers is given by Main[73].

Practolol *(17)* had to be withdrawn from the market because of serious side effects[62].

After it had been recognized that optimal β_1-selective receptor blockers could be synthesized by the use of certain substituents mainly in the 4-position of the phenyl nucleus, the terminal alkylamino group of the aliphatic side chain was also subjected to various modifications. The isopropanol chain as well as the secondary character of the alkylamino group had to remain unchanged. The molecular structure of toliprolol[32–34, 36], whose alkylamino group is linked with a substituted phenyl residue, can be regarded as a structural "lead".

Bevantolol[74–76] and in particular tolamolol[77–79], in which the alkylamino group contains a substituted phenoxy residue, produce a highly selective blockade of β_1-adrenoceptors. In combination with the aromatic phenyl ring, both the 3,4-dimethoxy substitution as well as the carbamoyl residue in the 4-position bring about the distinct transition to β_1-selectivity compared with the nonselective toliprolol.

This knowledge has led to intensive investigation of selectivity in wide-ranging fields of research. The molecular structures of the agents specified in Fig. 6 and 7 were of fundamental importance here.

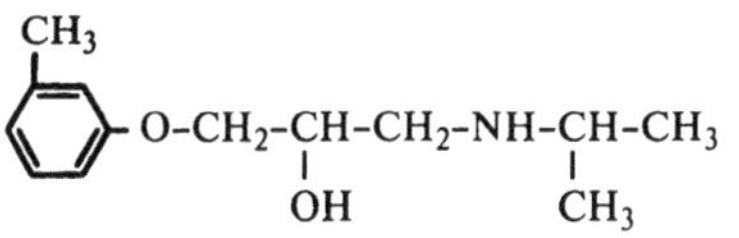

Toliprolol
Boehringer Ingelheim 21)

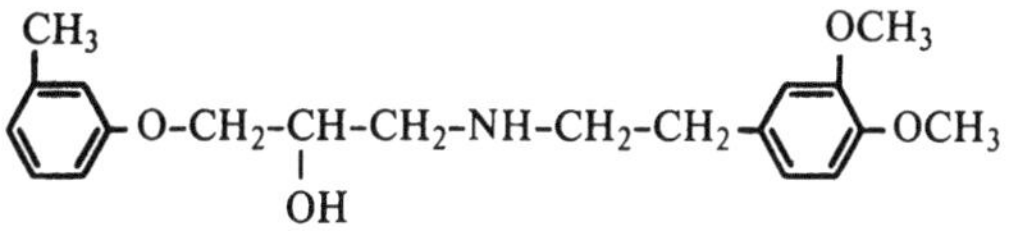

Bevantolol
Parke-Davis 22)

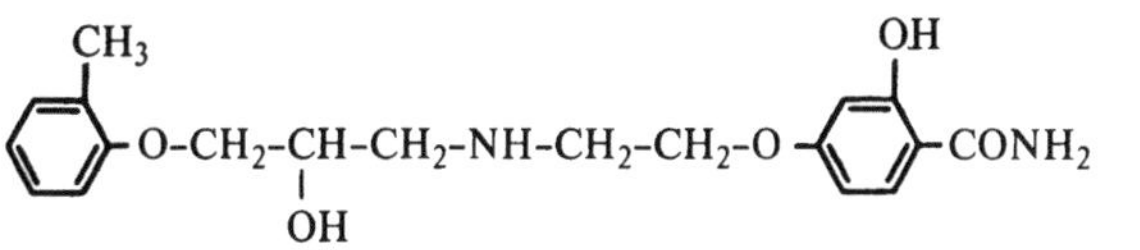

Tolamolol
Pfizer 23)

Fig. 7. Comparison of the structures of a nonselective β_1-adrenoceptor blocking drug and selective β_1-adrenoceptor blocking drugs

Starting from the structure of practolol and mostly using the companies' own β-blockers, substances with an acylamino group in the 4-position of the phenoxy residue were synthesized. A survey on newer types of compounds is presented in Fig. 8.

In the syntheses of compounds of formula *(24)*, the residue R_3 was particularly varied[80]. The phenethyl residue of bevantolol *(22)* served as a structural lead. It was linked to the amino group in the same way as in bevantolol. Since at the same time an acylamino group is present in the 4-position of the phenyl residue, a high β_1-selectivity was attained[80, 81]. New features are the double substitution of the phenyl residue with the acylamino group in the 4-position and a heterocycle in the 2-position[82, 83]. These agents summarized in formula *(25)* exhibit a marked

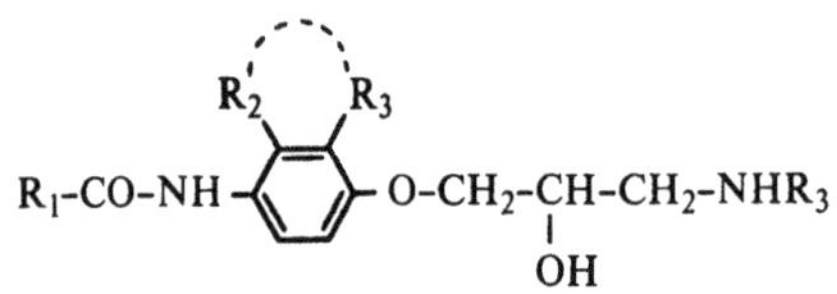

G. Washington Univ. 24)

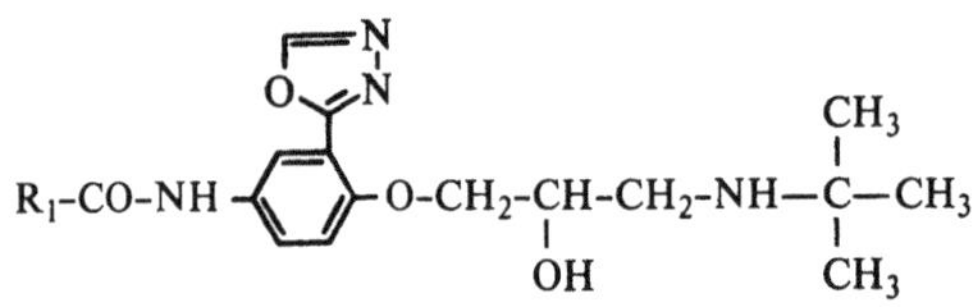

Beiersdorf 25)

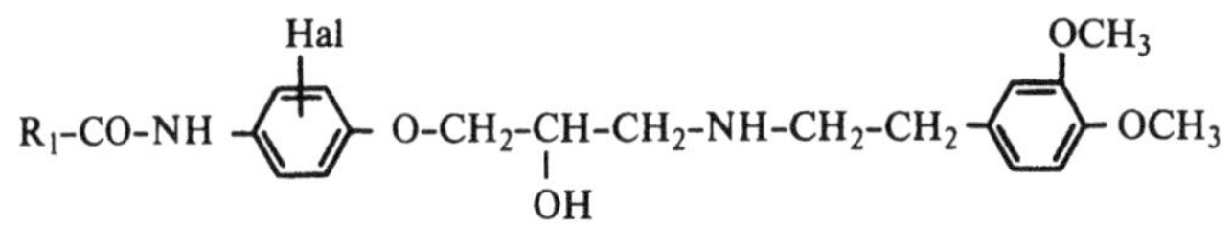

26)

Akad. d. Wiss., DDR

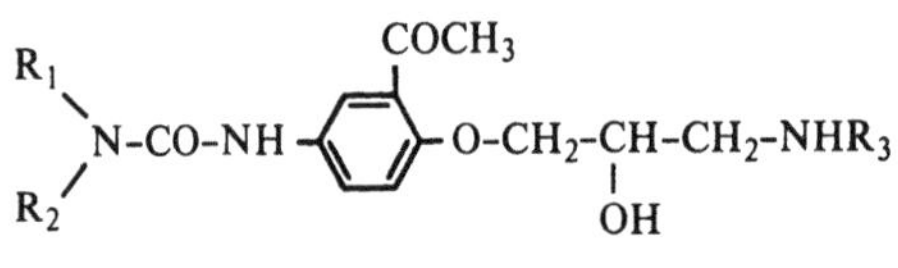

Chemie Linz 27)
Celiprolol
$R_1, R_2 = C_2H_5, R_3 = tC_4H_9$

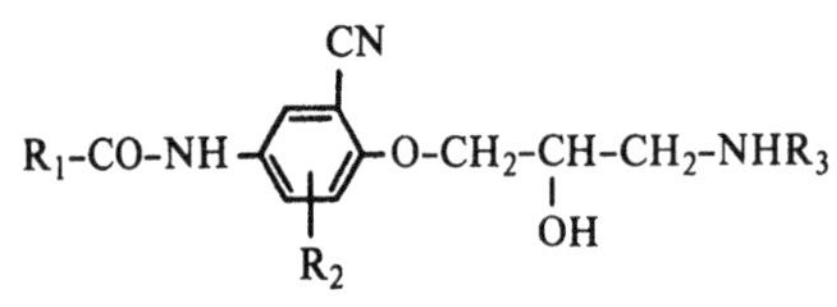

Boehringer Ingelheim 28)

Fig. 8. Selective β_1-adrenoceptor blocking drugs with acylamino substitution in the 4-position of the phenoxy residue

β_1-adrenoceptor blockade. Structures very closely related to those of class *(24)* compounds are to be discerned in formula *(26)* substances[84]. Their activity spectra are very similar. A variation of the acylamino function is represented in formula *(27)*. By acylation of the amino group with the carbamoyl residue, an urea group was generated in the 4-position[85]. Chemically, the acetyl group inserted in the 2-position brings the substance class close to acebutolol. Celiprolol was developed from this series of β_1-selective agents[86]. The introduction of an acylamino residue into the bunitrolol molecule[87] and its homologs[88–90] has resulted in formula *(28)* compounds. They combine the pronounced β-receptor blockade of bunitrolol with excellent β_1-selectivity[91]. Structure-activity relationships in a homologous series from a small section of class *(28)* compounds are shown in Fig. 9. It constitutes a continuation of earlier studies[91a].

In formula *(29)*, only the residue R_1 was varied[92–94]. The pharmacological results allow a comparison with data of β_1-selective adrenoceptor blocking drugs.

It was pointed out above that in β-blockers the alkyl substituent of the aliphatic secondary amino group must be branched in order to develop a strong receptor blockade. This applies particularly to unsubstituted alkyl residues. As shown in Fig. 9, this rule also applies to formula *(29)* compounds. From the n-propyl to the isopropyl and from the n-butyl to the secondary and tertiary butyl group, there is an abrupt increase in the isoprenaline antagonism by a factor of about ten. This rule also extends to the unsaturated alkyl residues, as can be seen from example 9 in Fig. 9. The substitution of the phenol group for the basically linked ethyl residue considerably enhances receptor blockade, as can be seen from example 8.

At the same time, all active isoproterenol antagonists shown in Fig. 9 are highly selective β_1-adrenoceptor blockers. This property is to be attributed solely to the characteristic acylamino group in the 4-position. If the acylamino residue is located in the 3-, 5- or 6-position of the phenyl group, β_1-selectivity is largely lost[95, 96]. Hence, these compounds and their data are not included in the Table.

The relationship between activity and structure is also evident in Fig. 10. Whereas the isobutyroylamino substituent remains unchanged in formula *(29)*, the acyl residue differs in formula *(30)*. The most effective alkylamino residues (isopropyl, tert-butyl and 3-methylbutinyl-1) are taken from Fig. 9, and are thus included in the variation.

This list confirms that, without exception, all substances are highly active isoprenaline antagonists due to the optimal substitution pattern of the amino group in connection with the 2-cyano substitution of the phenoxy residue.

If the acyl residue (R_2–CO–) is lengthened, dependence of antagonistic potency on increasing chain length can be inferred in certain cases only. This applies to all substances in columns three and four. On the other hand, it becomes evident from columns two, three and four that a major increase in the selectivity quotient SQ[91] is associated with increasing chain length of R_1. It attains a maximum with the hexanoylamino group. Further investigations have shown that neither an increase in chain length nor an exchange of the alkyl group (R_1) by cycloalkyl, aralkyl or aryl residue[95, 96] lead to an enhancement of activity or to an increase in selectivity.

When such β_1-selective receptor blockers act on the isoprenaline-stimulated rise of free fatty acids (FFA), it was observed that these selective agents already inhibit FFA release at low doses in various animal species, whereas the rise in lactate due to

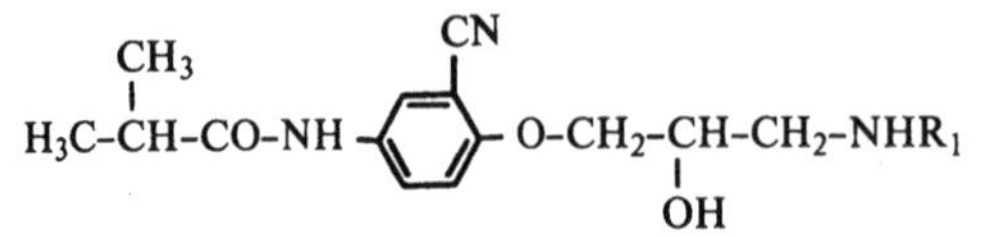

29)

Nr.	R_1	ED_{50}, mg/kg Guinea-pig, i.v.	Quotient of selectivity
1	CH_3	4,4	-
2	C_2H_5	0,53	-
3	nC_3H_7	1,05	-
4	$isoC_3H_7$	0,07	28
5	$n\text{-}C_4H_9$	2,3	-
6	$secC_4H_9$	0,11	34
7	$tertC_4H_9$	0,06	23
8	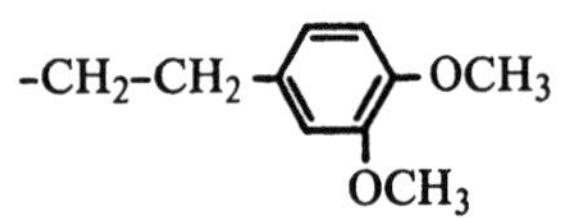	0,18	>96
9	$-C(CH_3)_2-C\equiv CH$ KOE 3290	0,03	32
10	Acebutolol	0,6	13
11	Atenolol	0,11	15
12	Metoprolol	0,04	16
13	Practolol	0,54	25

Fig. 9. β_1-adrenoceptor selectivity and isoproterenol antagonism depending on the substituent R_1

isoprenaline could only be blocked with high doses of the selective antagonist[97]. Non-β_1-selective receptor blockers (propranolol, bunitrolol) show hardly any differentiation between the inhibition of FFA release and lactate formation. With such β_1-selective receptor blockers, an attempt is being made to achieve additional advantages for the therapy of myocardial ischemia, in particular to attain an improvement in energy metabolism in the heart.

Whereas practolol was the structural lead for the agents shown in Figs. 8–10, with the goal of increasing the main effects and eliminating unacceptable side effects metoprolol was taken as the lead structure for further molecular variations.

Lengthening of the methoxyethyl substituent in the 4-position of the phenyl group of metoprolol promotes β_1-receptor selectivity[100] in formula *(32)* compounds. Its selectivity is especially enhanced by the terminal cyclopropyl group, which has an essential role in determining the activity spectrum of betaxolol[101–104]. Cicloprolol *(34)* differs from betaxolol *(33)* only in its second phenolether oxygen, their activity profiles showing common features[105, 106]. Original extensions of the side chain with the use of heteroatoms have led to the new class *(35)* and *(36)* compounds. They

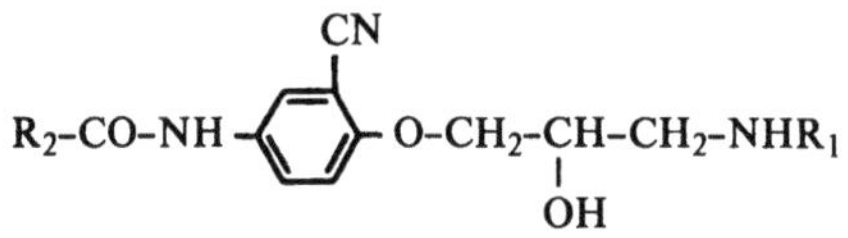

30)

$R_1 \rightarrow$	$-CH(CH_3)-CH_3$		$-C(CH_3)_2-CH_3$		$-C(CH_3)_2-C{\equiv}CH$	
$R_2 \downarrow$	ED_{50}, mg/kg GP*, i.v.	SQ** GP	ED_{50}, mg/kg GP, i.v.	SQ GP	ED_{50}, mg/kg GP, i.v.	SQ GP
CH_3	0,11	0	0,07	18	0,1	8
C_2H_5	0,02	13	0,04	21	0,06	11
nC_3H_7	0,26	8	0,04	0	0,02	17
$isoC_3H_7$	0,07	28	0,06	23	0,03	32
					(KOE 3290)	
n-C_4H_9	0,01	>24	0,1	0	0,08	12
tert.C_4H_9	0,11	38	0,04	25	0,09	61
nC_5H_{11}	0,02	336	0,09	100	0,02	195
			(KOE 4299)		(KOE 4302)	
nC_7H_{15}	0,03	67	0,04	96	0,1	81
Acebutolol	0,6	13				
Atenolol	0,11	15				

* GP = guinea-pig
** SQ = selectivity quotient

Fig. 10. β_1-adrenoceptor selectivity and isoproterenol antagonism depending on the substituents R_1 and R_2

have been synthesized independently in three research centers[107–109] and differ in their terminal groups and the substituents at the central phenyl group. They are potent β_1-selective adrenoceptor blockers[110] which might be developed in several indications.

In formula *(37)*, the phenyl residue of series *(34)* is replaced by a pyridyl residue. Similar β_1-selective properties to those in substance *(34)* are described, including antiarrhythmic effects[111].

The linking of the alkoxy group to the para-position of heterocycles[112, 113] represents an attempt to attain further differentiations of selectivity without reducing the pronounced β-blockade[114].

The very widely-based studies are documented by the selected examples of class *(38)* and *(39)* substances.

The special structural features of the β_1-selective β-blockers bevantolol and tolamolol were used as models in the synthesis of new β_1-selective agents. Typical examples are shown in Fig. 12.

Pacrinolol *(40)* is described as a specific β_1-sympatholytic agent without ISA[115], which has a distinct hypotensive effect in man[117]. The compound is substituted in

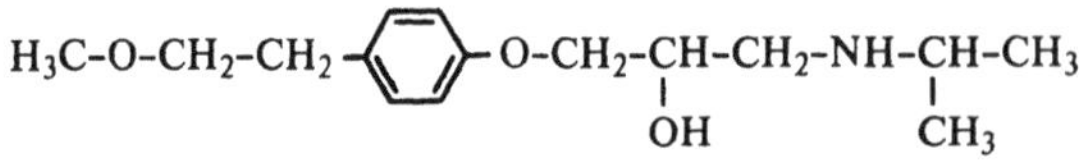
Metoprolol Ciba-Geigy 31)

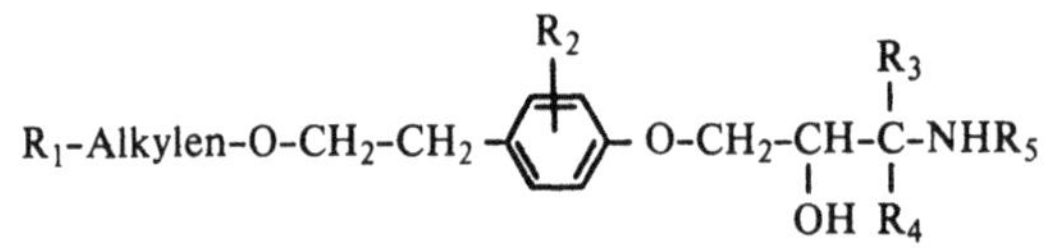
Beiersdorf 32)

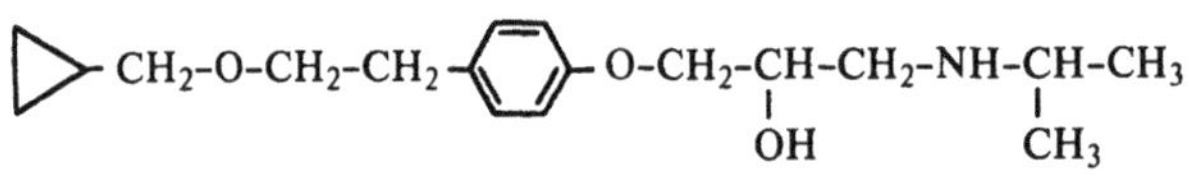
Betaxolol Synthelabo 33)

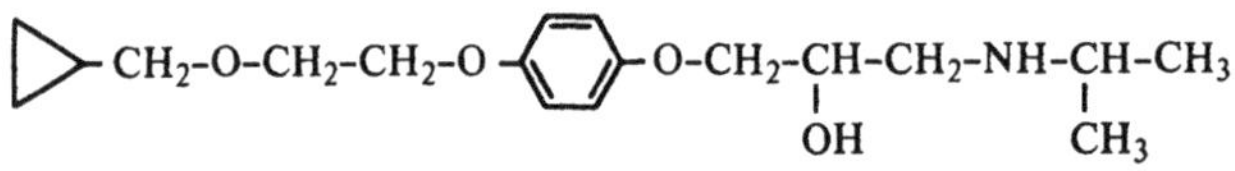
Cicloprolol Synthelabo 34)

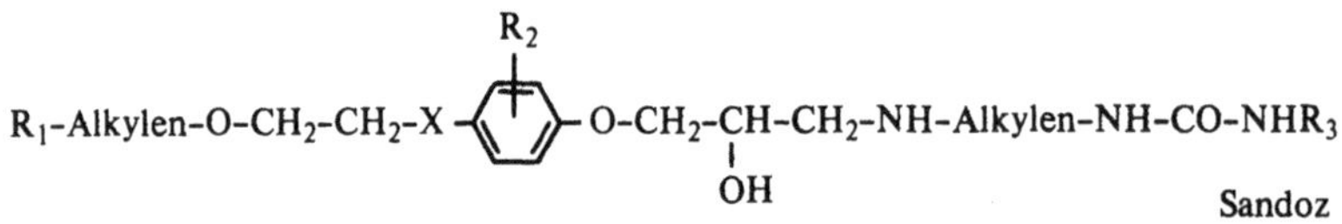
Sandoz 35)

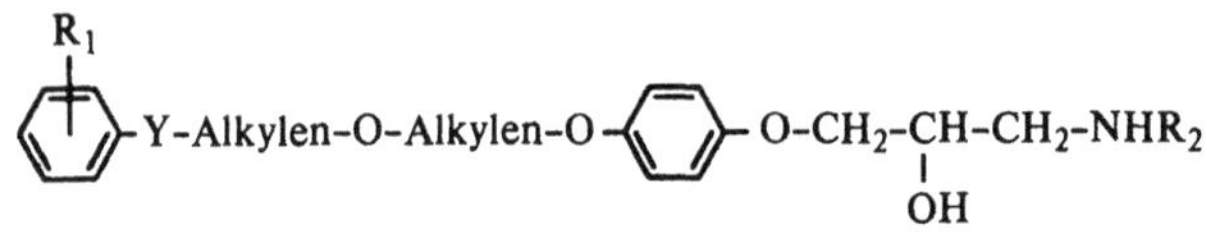
Hoffmann-La Roche 36)

(Ro 311118 = Flusoxolol)

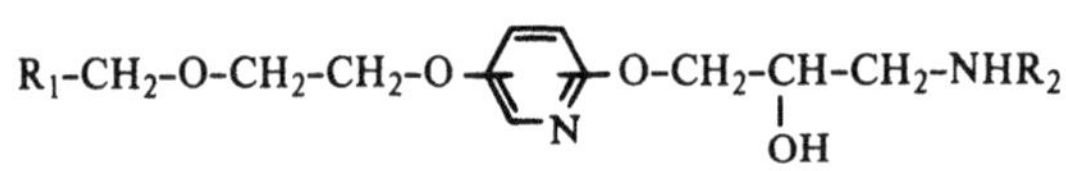
Synthelabo 37)

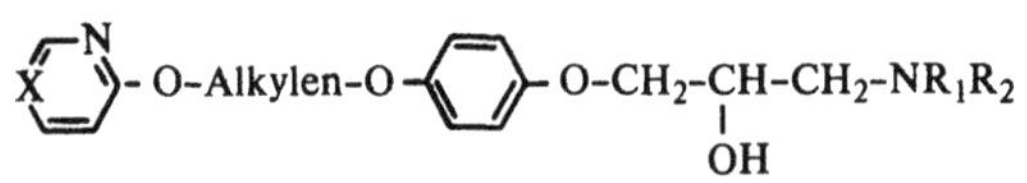
Yoshitomi 38)

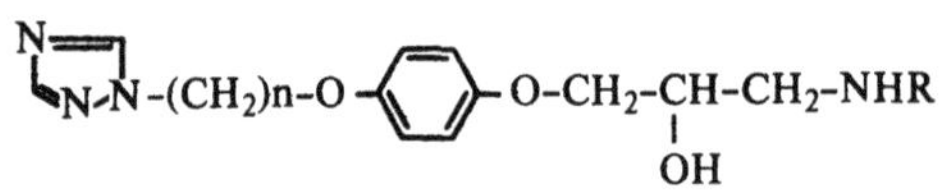
Hoffmann-La Roche 39)

Fig. 11. Selective β_1-adrenoceptor blocking drugs; congeners to metoprolol

the 4-position of the phenoxy residue in an original way with a 2-crotonic acid nitrile residue[116], and at the amino group with the 3,4-dimethoxyphenylethyl residue of bevantolol. The β_1-selective adrenoceptor blocker bometolol *(42)* has been derived from the substance class *(41)* developed in Japan[118, 119]. The antihypertensive effect of bometolol has contributed to further development[120]. The phenoxy residue of this agent shows in para-position the polar acetonyloxy residue, and at the same time is cyclized to a tetrahydroquinolinone. The S-enantiomer is responsible for β-blockade and selectivity[121]. In compound *(43)*, too, increased polarity is achieved by means of the imidazol residue in 4-position which, additionally, is combined with the

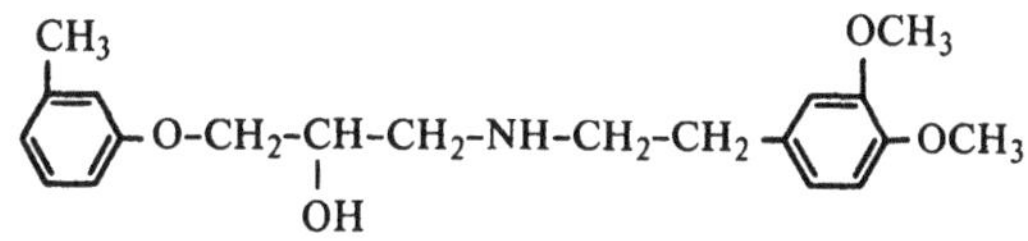

Bevantolol Parke-Davis 22)

S 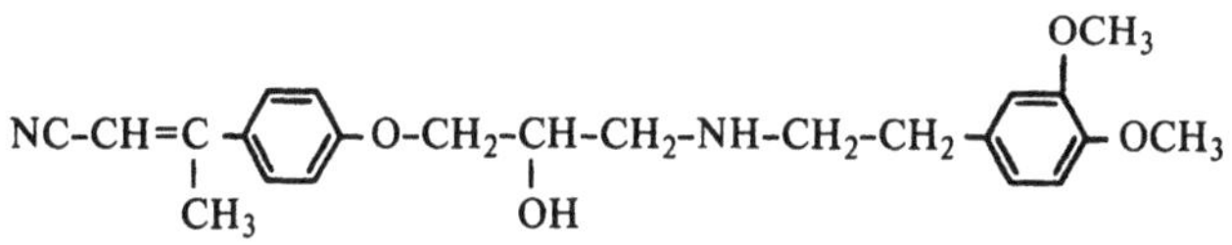

Pacrinolol Hoechst 40)

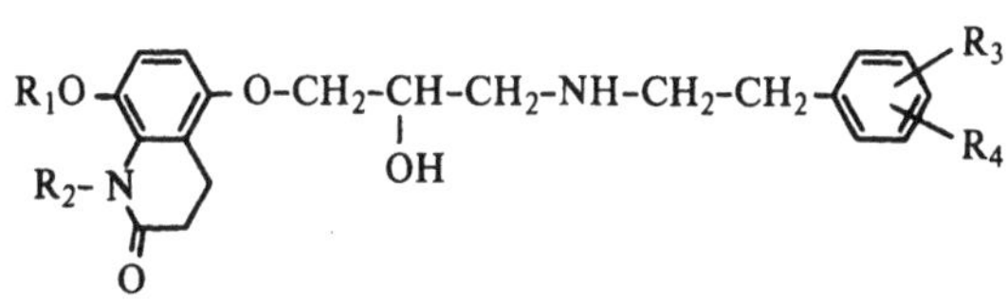

Otsuka 41)

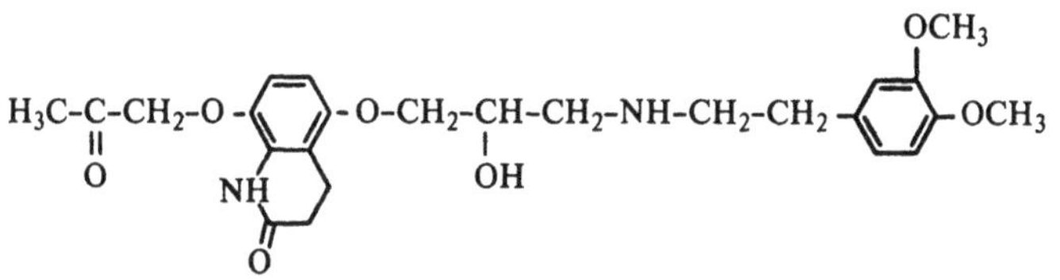

Bometolol Otsuka 42)

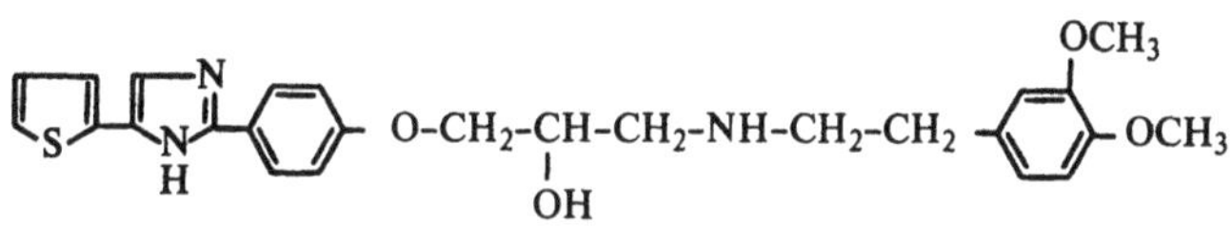

Merck 43)

S 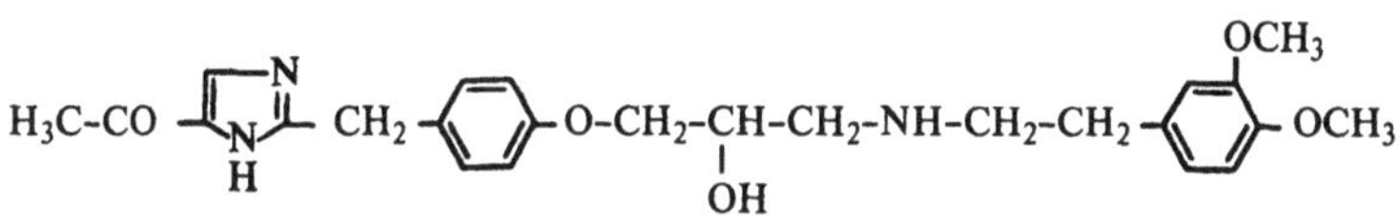

Merck 44)

Fig. 12. Selective β_1-adrenoceptor blocking drugs; congeners to bevantolol

thienyl group. The substance has a strong β_1-adrenoceptor affinity: the selectivity of the S-enantiomer exceeds that of atenolol by a factor of 100[122)]. The S-enantiomeric compound *(44)* also has once more a strongly polar substituent in the 4-position. This enantiomer is described as being extremely β_1-selective without ISA[123)]. These properties are attributed to the imidazolyl substituent. Due to the CH_2 bridge, there is an ideal distance from the phenyl residue, whereby the molecule appears optimally adapted to the receptor.

The molecular structure of the β_1-selective tolamolol *(23)* served as a lead for further syntheses. These yielded some new agents from which a selection will be presented below (Fig. 13).

The hydrophilic character of the compound class *(45)* results from a polyhydroxyalkyl residue R_1. These substances exert a β_1-selective effect in combination with an ISA[124)]. Compounds of formula *(46)*[125)] are closely related chemically and pharmacologically and are thus also adrenoceptor blockers with high β_1-selectivity.

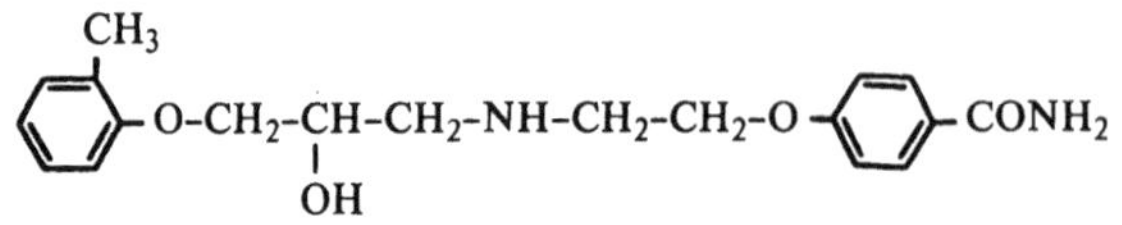

Tolamolol
Pfizer 23)

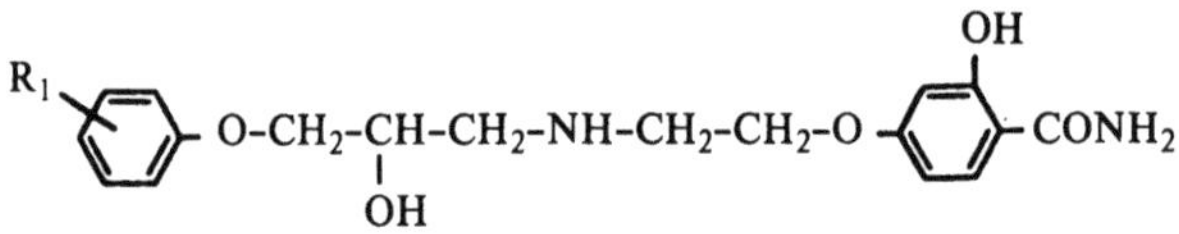

Ciba-Geigy 45)

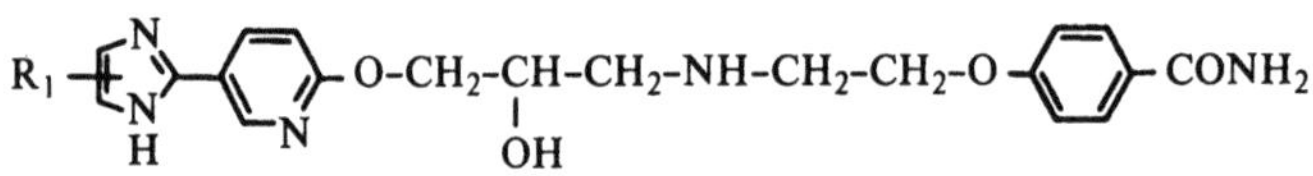

Ciba-Geigy 46)

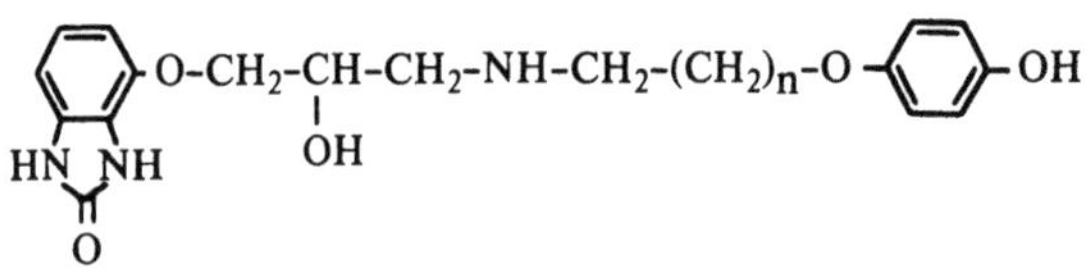

Ciba-Geigy 47)

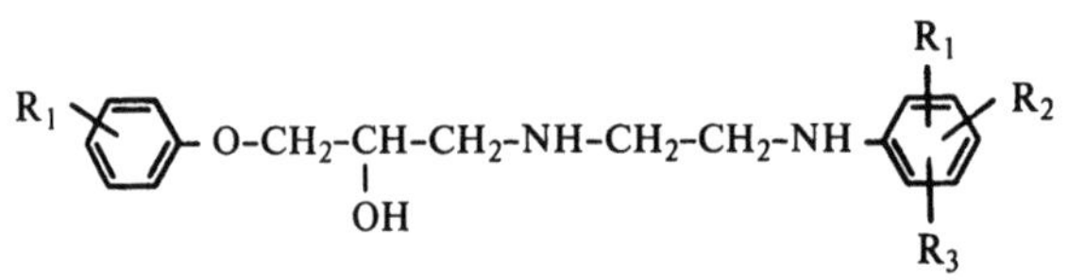

VEB Arzneimittelwerk
Dresden 48)

Fig. 13. Selective β_1-adrenoceptor blocking drugs; congeners to tolamolol

In contrast to these heterocyclically substituted substances from Ciba-Geigy, the isopropanolamine derivatives *(47),* also from Ciba-Geigy, are heterocyclically annelated[126)]. They elicit a long-lasting β_1-blockade. The long duration of action is essentially due to the heterocycle.

In formula *(48),* the phenoxyethyl residue of tolamolol is replaced by an anilinoethyl residue[127)]. In the action profile of this β-blocker series, marked ISA is present in addition to β_1-selectivity.

Besides these structures derived from the β_1-selective adrenoceptor blockers metoprolol, bevantolol and tolamolol, β_1-selective agents based on the structure of bunitrolol have also been found. There are also agents which can be fitted into the above-mentioned structures only with difficulty. Some of these are presented in Fig. 14.

The compounds *(49)* to *(52)* can be regarded structurally as being derived from bunitrolol[87)]. They are linked either directly with the alkylamino group or via an alkylen bridge which is interrupted by a heteroatom with a heterocycle or a phenyl group.

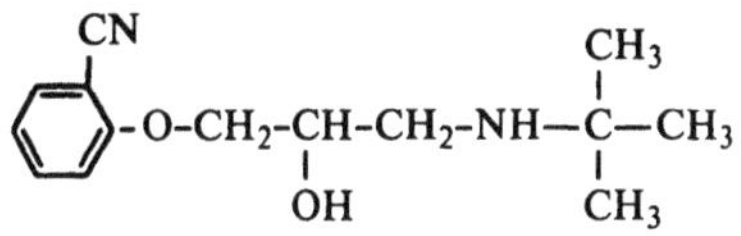

Bunitrolol
Boehringer Ingelheim 92)

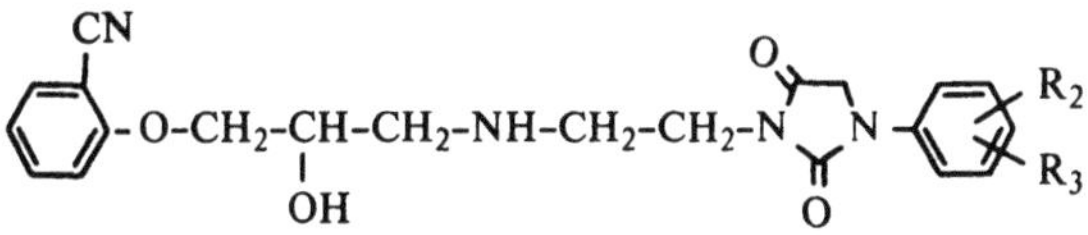

P O 160 A (R_2, R_3 = H)
Pierrel 49)

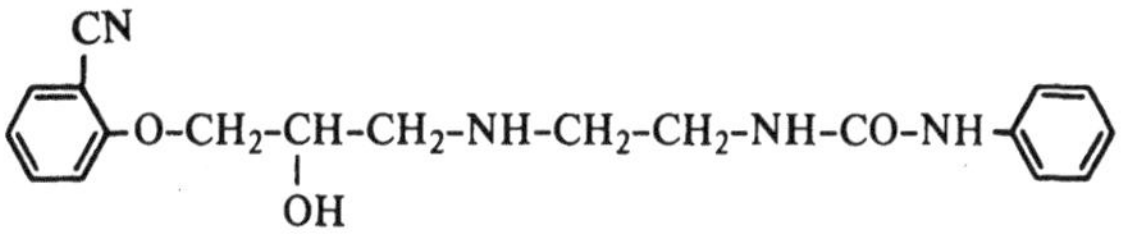

ICI 89406
ICI 50)

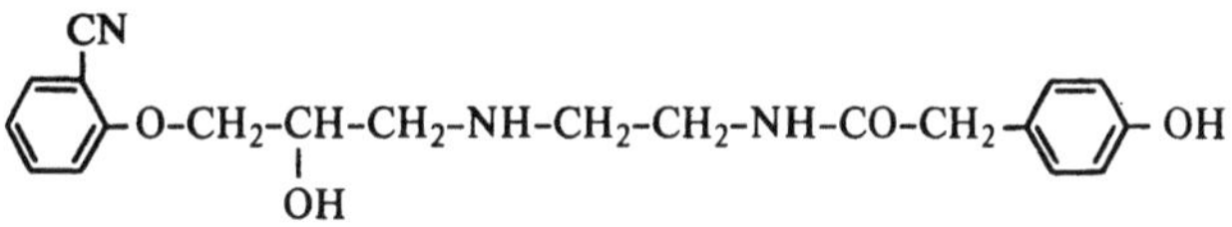

Epanolol
ICI 141292
ICI 51)

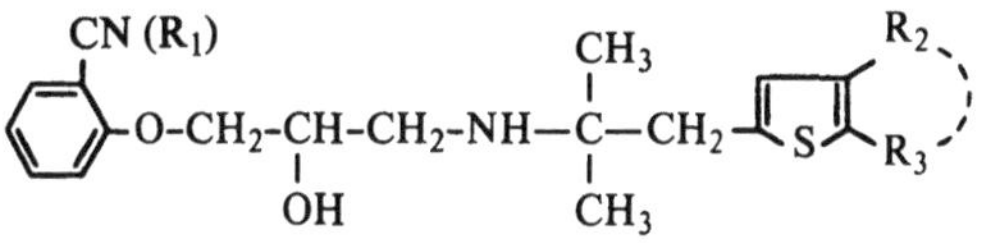

Bristol-Myers 52)

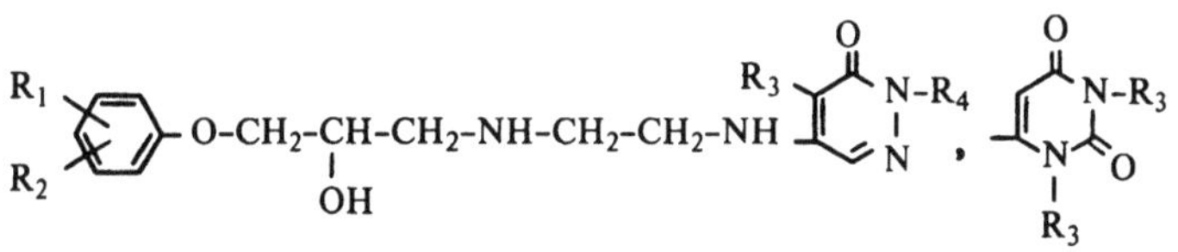

Cassella 53)

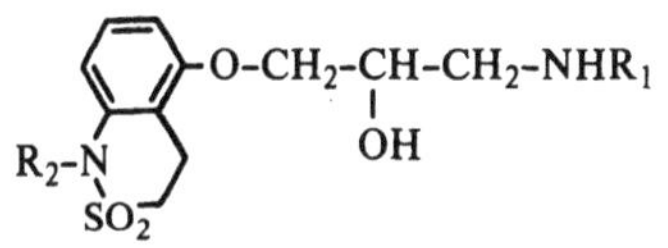

Suntory 54)

Fig. 14. Selective β_1-adrenoceptor blocking drugs; congeners to bunitrolol and others

Formula *(49)* substances are potent and distinctly β_1-selective adrenoceptor blockers[128, 129]. The substance P0160A under development is more selective than atenolol and has a weak ISA[130]. Effective β_1-adrenoceptor blockers are the closely related substances ICI 89406[131] *(50)* and ICI 141292 (epanolol) *(51)*[132], which display a high β_1-selectivity (also in man) with a long duration of action[133–135]. In addition to β_1-selectivity, the urea derivative ICI 89406 *(50)*[136] also has a marked ISA[137]. In compounds of the formulae *(52)* and *(53)*, which are likewise heterocyclically substituted, the selective β_1-adrenoceptor blockade is associated with an antihypertensive effect[138–140]. The novel benzothiazine derivatives *(54)* are reported to display high β_1-selectivity[141] and to be superior to reference substances (carteolol, acebutolol, propranolol).

These extensive studies show that it is not only the research divisions of the pharmaceutical industry that have attached and will continue to attach great importance to selective β_1-adrenoceptor blockade. Clinical investigations alone will show whether these novel chemical developments come up to medical expectations.

3.3 Enantiomeric β-Adrenoceptor Blocking Drugs

In Fig. 12, two substances are listed under formulae *(40)* and *(44)* which have been investigated in animals and gone into development as pure enantiomers with S-configuration[115–117, 123].

Their enantioselective synthesis or the optical resolution of the racemic compounds into enantiomerically pure drugs by fractional crystallization using optically active acids imposes especially high demands on chemical technology.

Figure 15 shows further β-adrenoceptor blockers which have either been introduced onto the market as drugs or which are still in the biological development stage.

All β-adrenoceptor blockers have at least one optically active center. In general, they are racemic compounds and used as such in therapy. According to the investigations by Howe and Shanks[142], in those mixtures of S- and R-enantiomers the S-enantiomer is the biologically active component.

S 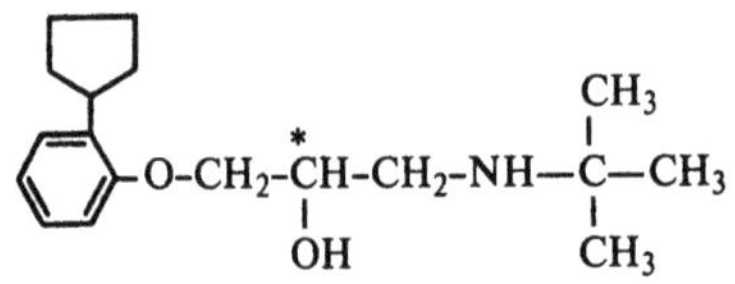

Penbutolol Hoechst 55)

S 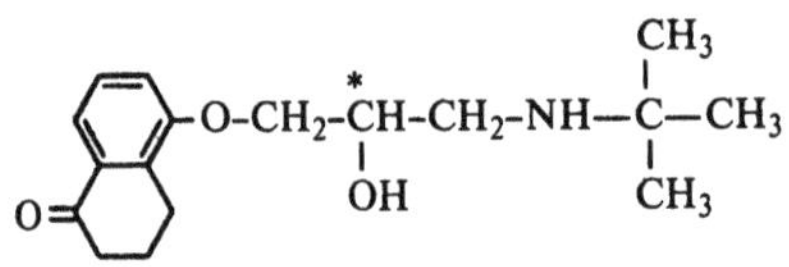

Levobunolol Warner-Lambert 56)

S 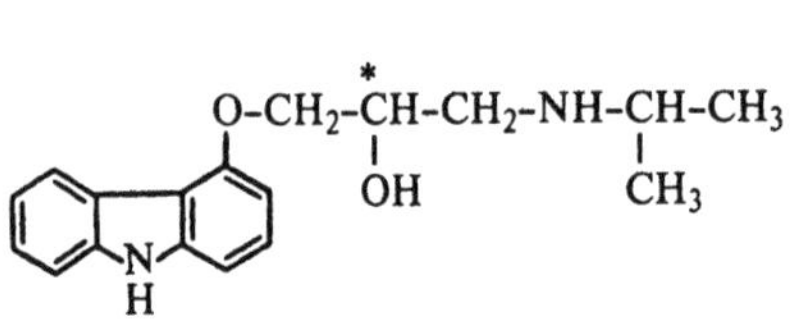

Carazolol Boehringer Mannheim 57)

S 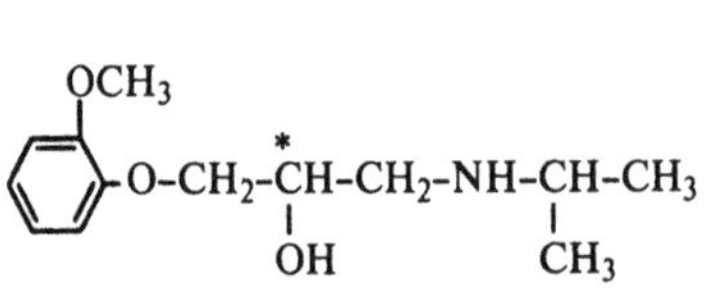

Moprolol Simes 58)

* = chiral center

Fig. 15. S-Enantiomer beta adrenoceptor blocking drugs

The use of the pure enantiomer exerting the clinical pharmacological activity has the advantage that its therapeutically inactive counterpart is absent and thus cannot produce any adverse effects. However, such considerations are hardly likely to play any role in the case of highly active and thus low-dosed drugs.

The β-adrenoceptor blockers in formulae *(55)* to *(58)* mentioned in Fig. 15 are not β_1-selective agents. Their indications are mainly in the areas of coronary heart disease, vascular hypertension and glaucoma.

Penbutolol *(55)*[143] was the first enantiomerically pure β-adrenoceptor blocker to be introduced into therapy. Therapeutic effects were attained both in the treatment of hypertension and angina[144–146]. Starting from racemic bunolol, it was possible to achieve optical resolution of the S- and R-enantiomers[147–149]. The levobunolol *(56)* is also used in the area of hypertension and angina as well as glaucoma[150–152]. Because of a lower rate of metabolism, the S-form has a longer halflife than the R-enantiomer. The optical resolution of carazolol[153] and enantioselective synthesis[154] have led to the S- and R-enantiomers *(57)*. Their properties are measured in comparison with carazolol[155, 156]. Starting from early synthesized β-adrenoceptor blockers[38], optical resolution yielded S-moprolol *(58)* which, in contrast to its R-form, has therapeutic properties[157]. The reduction of peripheral resistance appears to contribute to hypotension[158, 159].

3.4 Short-Acting β-Adrenoceptor Blocking Drugs

The synthesis and development of short and ultrashort-acting β-adrenoceptor blockers is based on the notion that such substances may afford advantages in certain clinical situations. I. v. infusion allows the desired blood level to be reached rapidly[160], but also allows a rapid adjustment of this level as required by the changing adrenergic status accompanying the dynamic character of an infarct in the acute stage[161]. After infusion, the β-blocker effect and any unexpected or undesired side effects subside rapidly, and a possible cardiodepression is also prevented. Possible problems in the airways are slight, as is also the case for the development of coronary vasospasm[162, 163]. Such β-adrenoceptor blockers are also useful for a therapeutic test when the patient has to be treated with a long-acting β-blocker. Patient reaction and tolerance to the product can then be recognized very quickly[164]. In connection with surgical operations as well as after infarction, the use of ultrashort-acting β-adrenoceptor blockers is recommended[165].

The following Fig. 16 provides a survey of some chemical developments of short-acting β-adrenoceptor blockers.

The substance esmolol *(59)* under development and all substance classes *(60)* to *(63)* have an ester group in common[161, 166–169]. This ensures rapid hydrolytic degradation. This was observed not only in animals, but also during investigations in man[164, 170]. In compounds substituted in the 4-position, a β_1-selectivity is observed[171]. Compounds of this kind have also been investigated for the treatment of glaucoma[167].

The short-acting β-adrenoceptor blockers are still largely in the development stage.

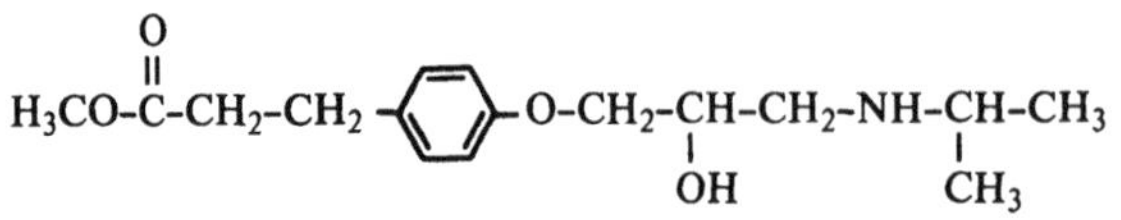

Esmolol
Cyanamid
59)

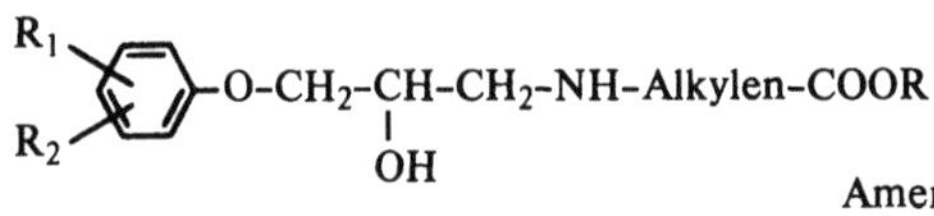

60)

Americ. Hosp. Suppl. Co.

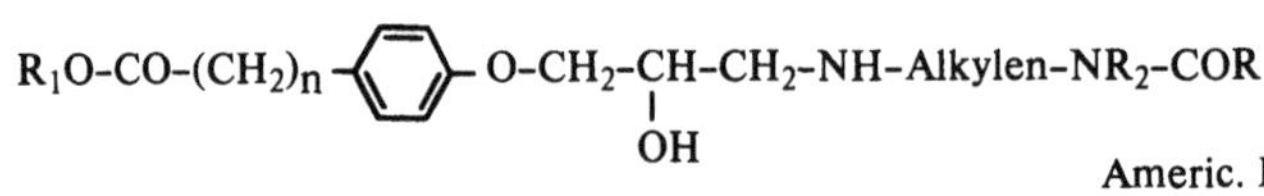

61)

Americ. Hosp. Suppl. Co.

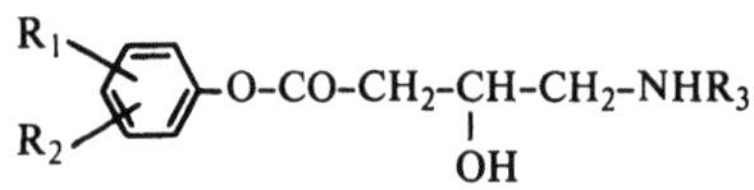

62)

Americ. Hosp. Suppl. Co.

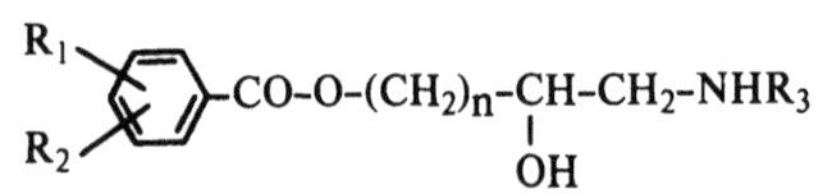

63)

Americ. Hosp. Suppl. Co.

Fig. 16. Ultrashort-acting beta adrenoceptor blocking drugs

4 β-Adrenoceptor Blocking Drugs as Antihypertensives

The synthesis of 1-alkylamino-3-phenoxy-2-propanols and the discovery of their β-adrenoceptor blocking properties (Köppe, Engelhardt, 1960/61)[28, 30] and their low toxicity led very quickly to the notion in various research centers, independently of one another, that they could be used for treatment of angina pectoris[28, 31, 172]. This was not a simple conceptual step: it was derived in part from a quite different area of work[28, 30], but very soon assumed practical forms and has been developed into an indispensable principle of therapy. At that time, the idea of treating vascular hypertension with these agents as a second indication did not occur. Moreover, the possibility of using them was not foreseeable.

The observation by Prichard[173, 174] was therefore a surprise: he established that such a new drug when used in the treatment of angina patients with hypertension brings about a marked reduction in blood pressure after a certain latency period.

The antihypertensive effect detected relatively early (1964) was only slowly exploited in the treatment of hypertension[8, 175)]. Although numerous theories to explain the mode of action of these drugs as antihypertensives have been postulated, none of them has been completely accepted to date. It is suspected that the antihypertensive action of the β-adrenoceptor blockers is not based on only one mechanism, but that this is a complex process attributable to diverse sites of action. Action on the renin-angiotensin system associated with a reduction in angiotensin II formation, reduction in cardiac output, an influence on baroreceptors as well as on central adrenergic receptors are discussed[8, 175)].

All β-adrenoceptor blockers exert an antihypertensive effect[176, 177)] which is independent of ISA, β_1-selectivity or membrane-stabilizing action. In 40 to 60% of the patients, a normalization of blood pressure can be achieved. In particular, younger patients respond to monotherapy with β-adrenoceptor blockers[177)]. Further

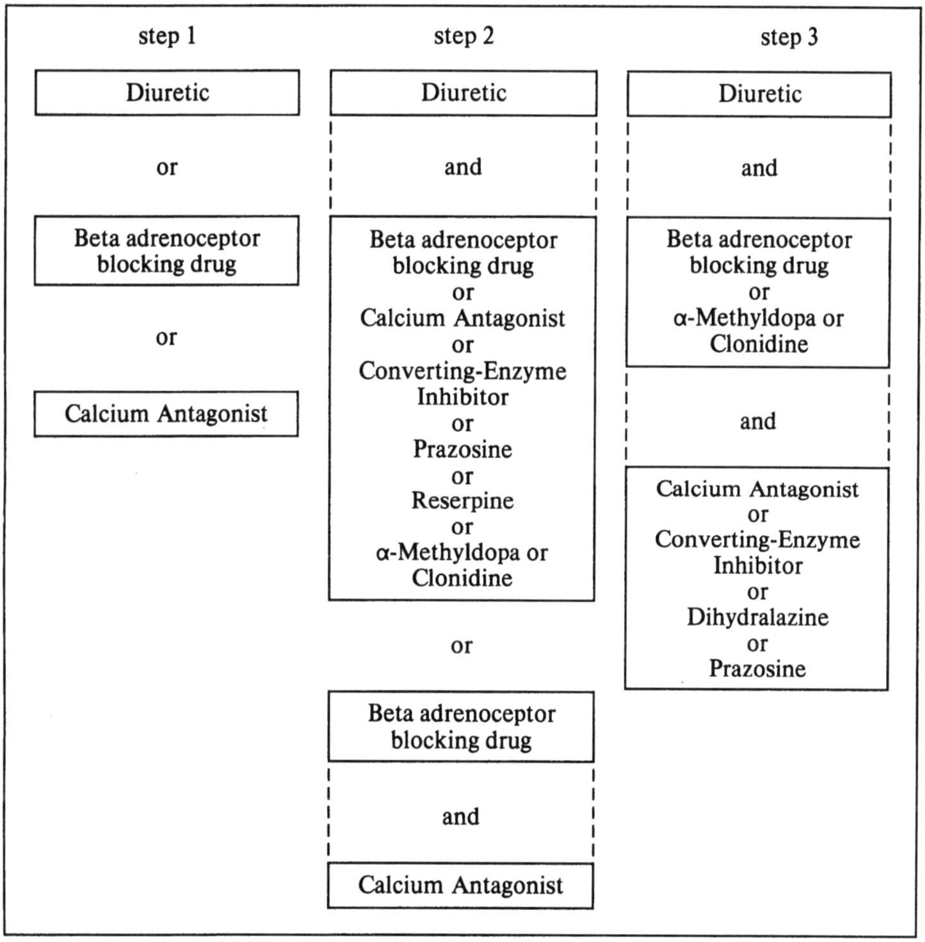

Fig. 17. Plan of antihypertensive drug therapy, recommended by the German League for the Control of High Blood Pressure

logical indications are mild hypertension and raised blood pressure with a high pulse rate. For the treatment of high blood pressure, phase plans have been worked out by national specialist committees and by the WHO. They contain recommendations for a stepwise procedure in the administration of antihypertensives[178].

It can be seen from Fig. 17 that β-adrenoceptor blockers are recommended in treatment of hypertension in the first stage of treatment as monosubstances, in the second treatment stage in combination with a diuretic and in the third stage in combination with a diuretic plus a vasodilator in severe forms of hypertensive disease. The table clearly shows that they are relatively mild drugs whose hypotensive effect can be intensified by combination with further antihypertensives.

Administration of such fixed combinations has proved advantageous. It was possible to raise success rates from about 40 to 60% to about 80 to 90% as compared to the monosubstances. These combination preparations available on the market will not be dealt with in further detail here.

It is evident that, using structural variations, research activities have been directed towards the development of β-adrenoceptor blockers. These are aimed at bringing about an intensified hypotensive effect with the same therapeutic ratio and the same or decreased adverse effects. Their duration of action should be prolonged if possible. Efforts are also under way to combine, in one and the same molecule, the activity profile of a fixed combination of a β-adrenoceptor blocker plus a diuretic or the effective activity spectrum of a fixed combination of a β-adrenoceptor blocker and a vasodilator.

A selection of such research activities going on all over the world is presented below. The sequence follows the phase plan of hypertension treatment.

4.1 *Enhancement of Antihypertensive Action by Molecular Variations*

Variations in the structure of known β-adrenoceptor blockers, which are used as monosubstances in hypertension, aim at increasing the antihypertensive effect and prolonging the duration of action. The duration of action can be influenced for example by hydrophilic substituents or new molecular structures. It is known that lipophilic β-adrenoceptor blockers are mainly eliminated via the liver and have a relatively short halflife. On the other hand, the duration of biological activity is very much longer, so that in the cardiac area (anti-angina preparations) a two to three times longer halflife can be expected. The antihypertensive effect of the same agents lasts even longer. It is not correlated with the blood levels and is still present even when β-blocker concentrations can no longer be detected[179]. On the other hand, hydrophilic β-adrenoceptor blockers, which generally have a longer halflife, are mainly excreted via the kidneys.

This information has been used in the synthesis of new β-blocker drugs. With regard to proven β-blockers, their essential structural characteristics have been preserved. The molecular structure has been extended at specific sites by characteristic components or linked with original components via bridges. In Fig. 18, compounds are presented which have been built up from the typical 1-amino-3-phenoxy-2-propanol skeleton, an alkylene bridge and a heterocycle.

From the general formula *(64)*[180], the compound *(65)* has been taken. Here, essential parts of atenolol are linked with the benzamidazolyl residue by an alkylene bridge. Besides a marked reduction in blood pressure, anti-arrhythmic effects have also been observed in this substance class[180]. The substance MEN 935 under development belongs to compound class *(66)*[181]. Here, the propranolol skeleton has been linked with the benzimidazolonyl residue. Due to its slow metabolization, this β-adrenolytic produces a long-lasting antihypertensive effect which is intensified by a

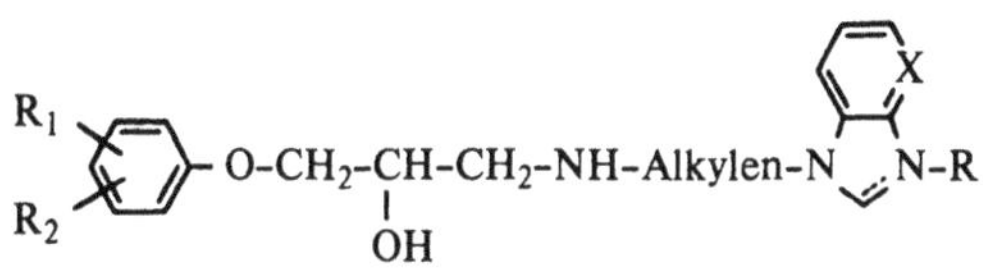

Berlex 64)

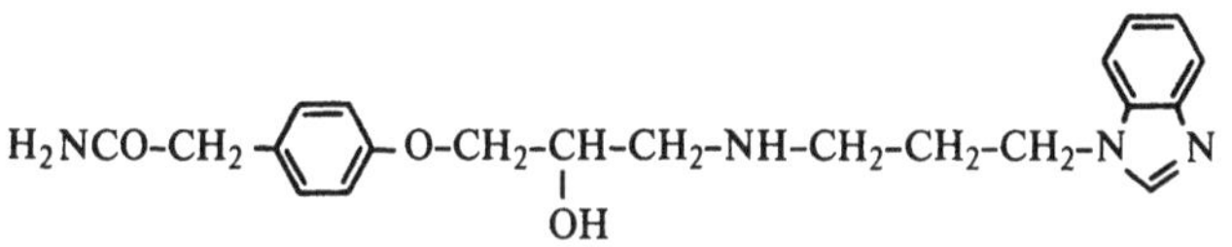

Berlex 65)

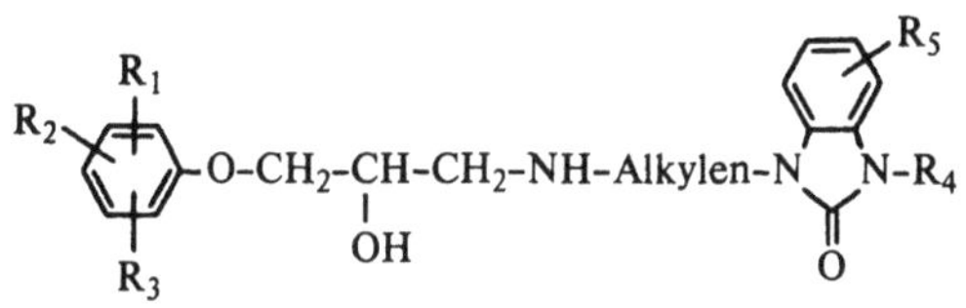

Boehringer Ingelheim 66)

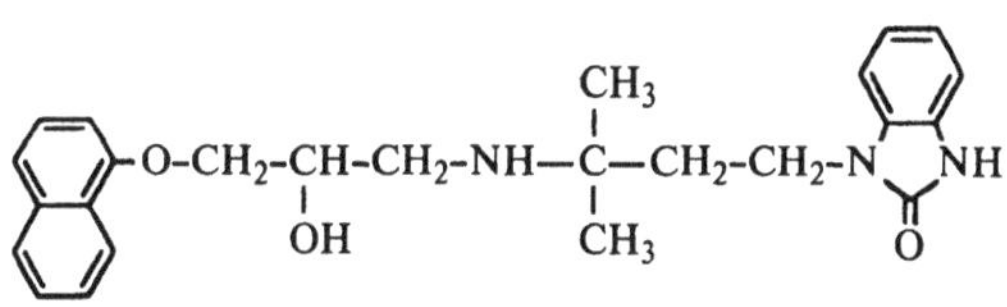

Adimolol (MEN 935) 67)
Boehringer Ingelheim

S

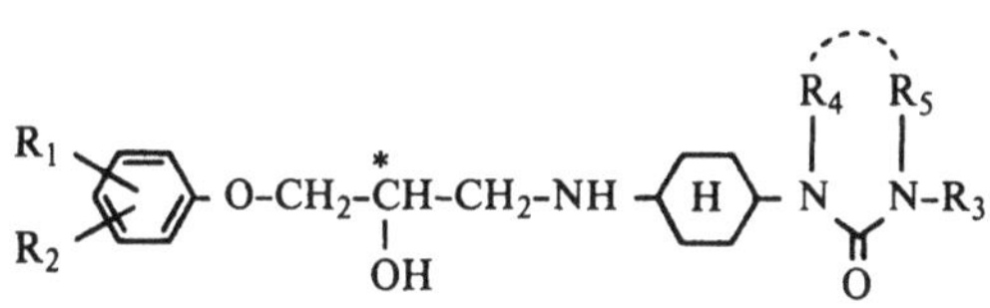

Ciba-Geigy 68)

S

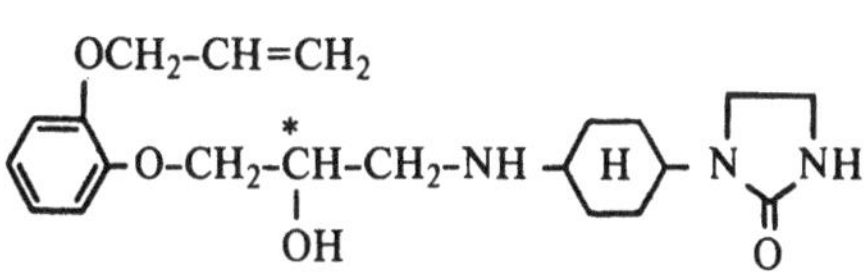

Ciba-Geigy 69)

Fig. 18. Antihypertensive beta adrenoceptor blocking drugs by heterocyclic substitution of the alkylamino group

mild α-adrenergic blocking effect[182–184]. Formula *(68)* compounds display a similar activity spectrum. They act above all as isoprenaline antagonists in which the hypotensive effect is intensified by additional α-adrenergic blockade[185]. Regarding these substances (especially compound *(69)*), it is of interest that they were synthesized from oxirans of enantiomerically uniform configuration and cyclohexylene derivatives substituted in the cis-position. They are thus present in a biologically highly active form. These antihypertensive substances do not exert any orthostatic effects.

If 1-aryloxymethyloxirans are reacted with basic heterocycles, compounds are formed of which a selection is shown in Fig. 19.

In formula *(70)*, the alkylamino residue of the β-adrenoceptor blocker has been replaced by a purine residue. This salient molecular alteration contributes to the distinct antihypertensive effect of these compounds[186]. In the compound classes *(71)* and *(72)*, the otherwise usual secondary alkylamino group of the β-blocker has been

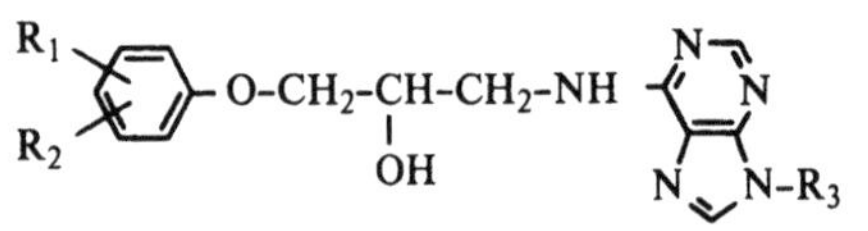

Searle 70)

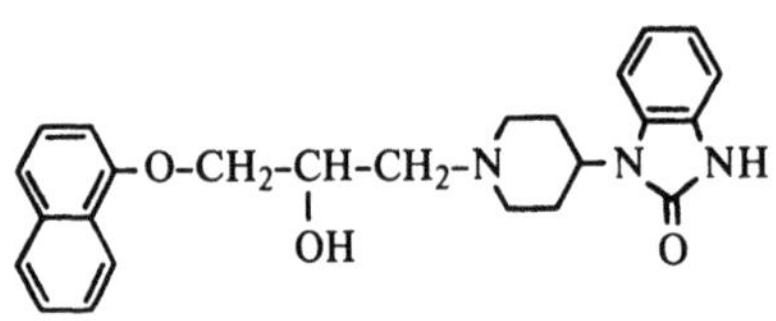

Kyowa 71)

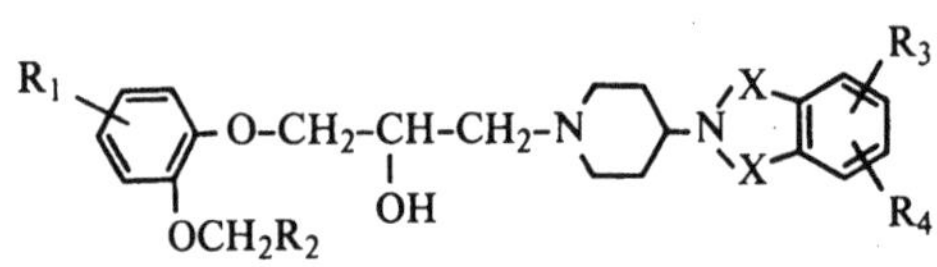

Wyeth 72)

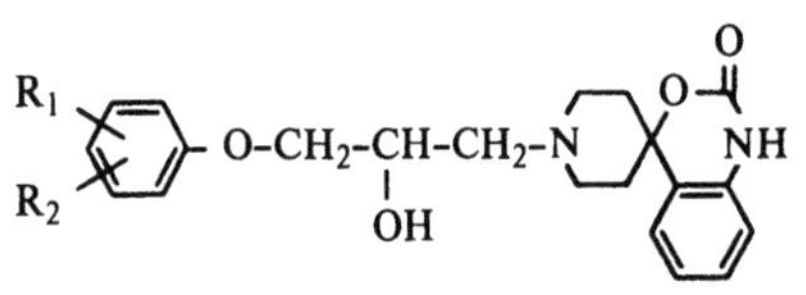

Syntex 73)

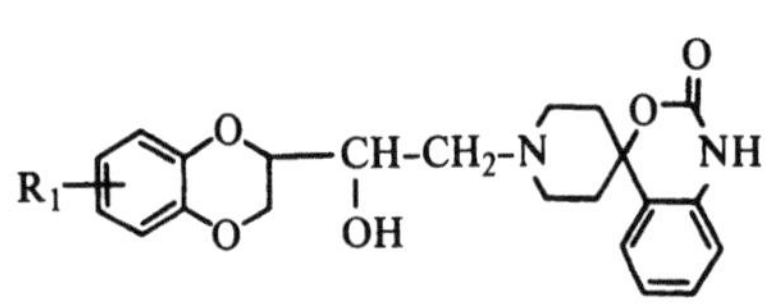

Syntex 74)

Fig. 19. Antihypertensive beta adrenoceptor blocking drugs by heterocyclic modification of the alkylamino group

replaced by a tertiary piperidino residue. This constitutes the bridge to the terminal heterocycles in the compound classes *(73)* and *(74)*. In animal studies in spontaneous hypertensive rats (SHR), the compounds *(71)* as well as *(73)* and *(74)* are antihypertensives[187, 188]. For substances *(73)* and *(74)*, a central effect has also been discussed in addition to the peripheral action. Compounds in formula *(74)* elicit a clear α_1-blockade in the anesthetized dog[188].

4.2 Enhancement of Antihypertensive Action by Additional α-Blockade

Figure 18 shows compounds in which a blockade of α-adrenoceptors was observed besides the blockade of β-adrenoceptors[182–185]. The former may contribute to the increase in the antihypertensive effect.

A simultaneous blockade of β- and α-receptors was observed for the first time with the phenethanolamine derivative labetalol[189–191]. This drug brings about a comparatively strong, nonselective β-adrenoceptor blockade and a weaker, selective blockade of α_1-adrenoceptors[192, 193]. At the same time, a membrane-stabilizing effect is observed. As a result, there is a lowering of blood pressure which is plausible: α-blockade lowers the peripheral resistance to flow. At the same time, cardiac output is reduced and the renin-angiotensin system may be influenced by the β-blockade. The reflex tachycardia brought about by α-blockade is abolished by the β-blockade[194, 195].

Of course, there have been efforts to improve this action profile in the synthesis of new β-adrenoceptor blockers. The goal of such research is to obtain substances with a balanced α/β-blockade without adverse effects. The chemical structure of labetalol served as a lead in the synthesis of the substances *(76)* and *(77)* shown below.

When the structures in Fig. 20 are compared, it is noticeable that besides labetalol there is only one further phenethanolamine derivative (medroxalol) present. It differs structurally from labetalol only in that it has an additional methylenedioxy substitution. It also becomes evident in the structural field of the α/β-adrenoceptor blockers, that in parallel to the purely β-adrenoceptor blockers from the 1-alkylamino-3-phenoxy-2-propanol class, a phenethanolamine derivative possesses an interesting activity profile by way of exception only.

The administration of medroxalol to patients with mild hypertension led to normal blood pressure and normal heart rate[196, 197]. It was well tolerated, but slight orthostatic effects were observed. The four stereoisomers are similar in their activity profile[198]. The substance under development, KF-4317 *(77)*, has an activity spectrum[199] similar to that of labetalol. It brings about a combined α/β-blockade. The molecular structure is derived from that of labetalol and atenolol[200]. Compounds of formulae *(78)* to *(80)* are to be regarded as derivatives of pindolol *(97)*. For these compounds, a joint α/β-blockade has been described[201–203]. Infendolol is under clinical development[202]. In healthy volunteers, α_2-blockade by the drug is more pronounced than α_1-blockade. Formulae *(81)* and *(78)* compounds differ merely in

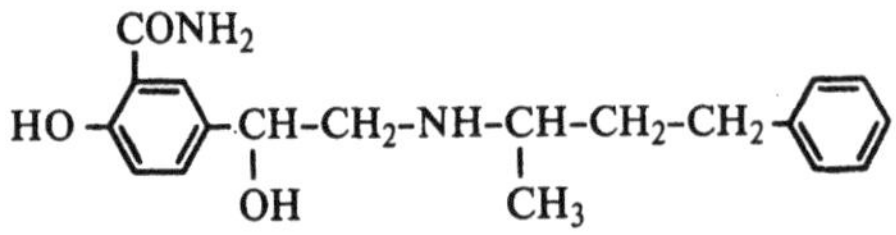

Labetalol
Allen + Hanburys 75)

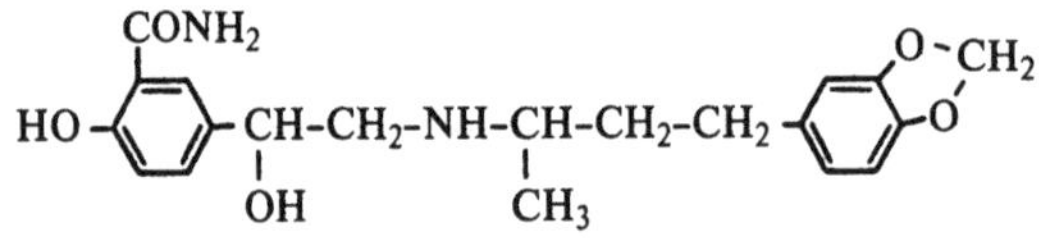

Medroxalol
Richardson-Merrell 76)
(RMI 81968)

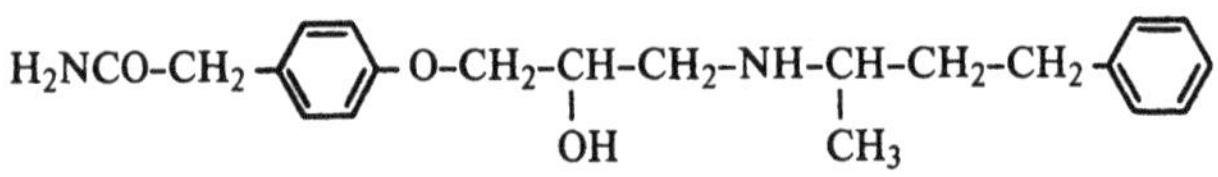

KF 4317
Kyowa 77)

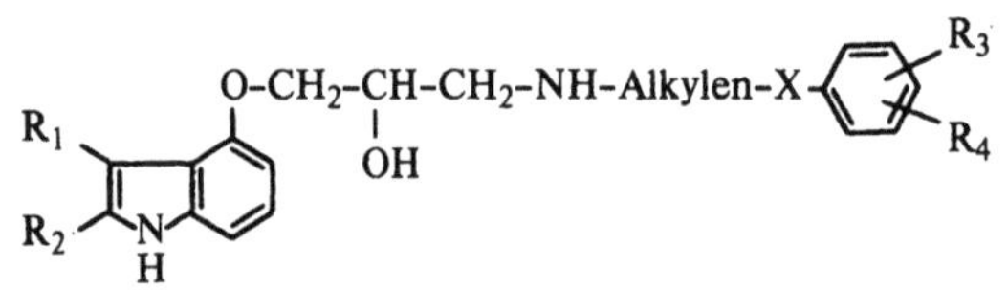

Sandoz 78)

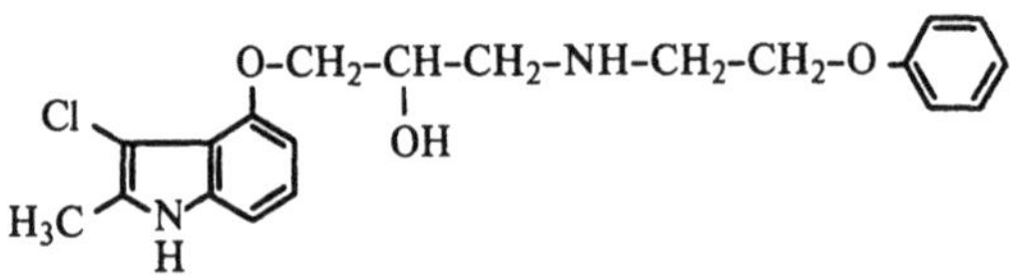

Infendolol
Sandoz 79)

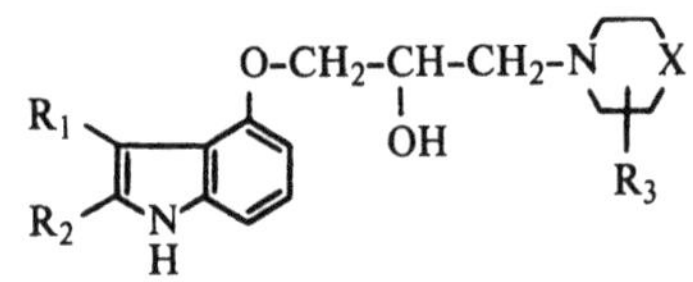

Sandoz 80)

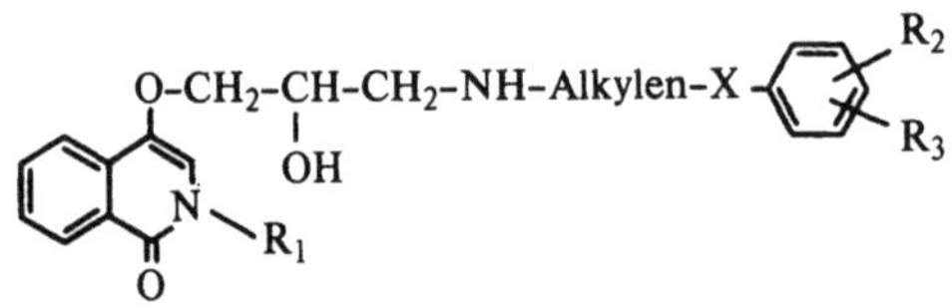

Nisshin 81)

Fig. 20. Antihypertensive beta adrenoceptor blocking drugs with alpha adrenoceptor blockade

the structure of the heterocycle and in the position of the ether bridge. The activity spectra are correspondingly similar due to the α/β-effects[204].

4.3 Enhancement of Antihypertensive Action by Additional Diuretic Properties

The German League for Control of Hypertension and the WHO recommend that the pharmacotherapy of hypertension should be approached step by step. In accordance with the phase plan, treatment should always be commenced with monotherapy[205]. Diuretics, β-adrenoceptor blockers and calcium antagonists are available as alternatives. If monotherapy with a β-blocker does not yield adequate results in the first phase, the combination of a β-blocker with a diuretic is suggested as a combination of interest for the second phase of therapy[205] (see Fig. 17). The medication can be given in a free or fixed combination. In order to administer as small an

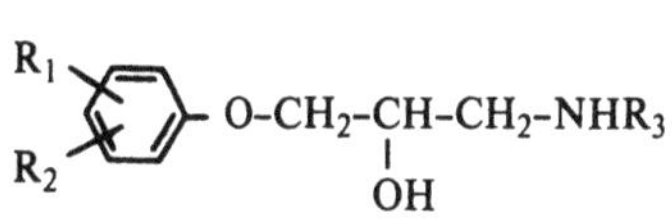

82)

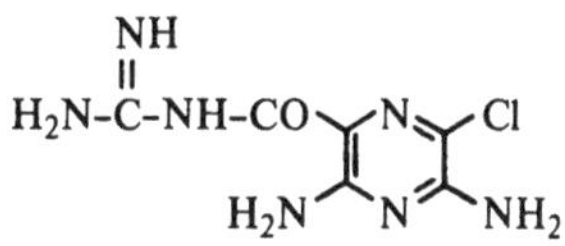

83)

Beta receptor blocking drug

Amiloride
(Diuretic)

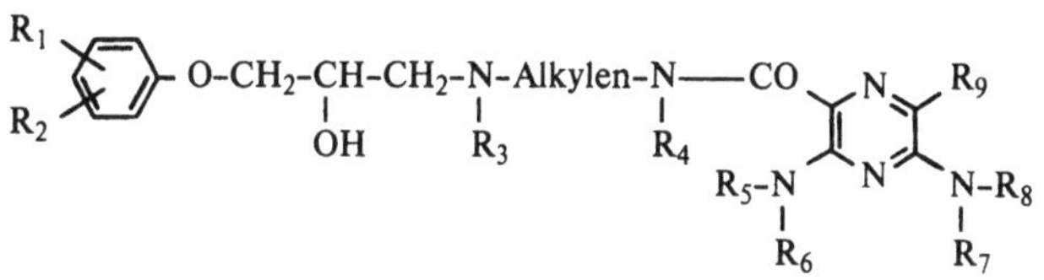

ICI 84)

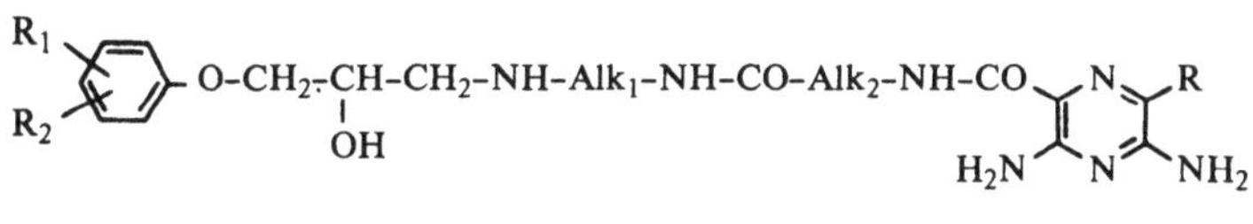

ICI 85)

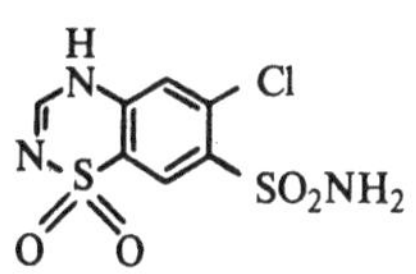

Clorothiazide
(Diuretic)

86)

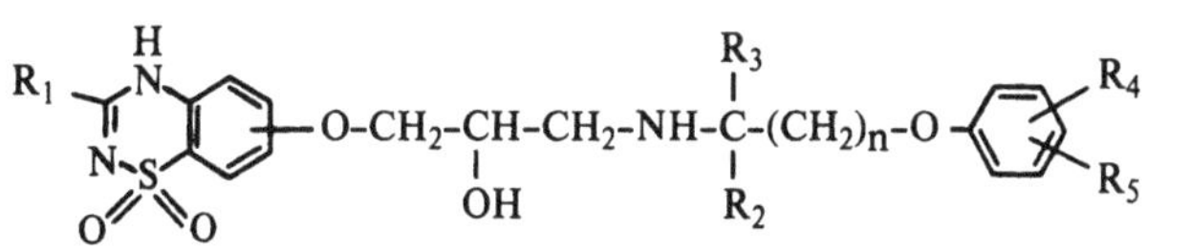

Merck 87)

Fig. 21. Antihypertensive beta adrenoceptor blocking drugs by structural combination with diuretics

amount of drug as possible, chemical research is trying out new approaches here. These aim at the combination of β-adrenoceptor blockade and diuresis in one and the same molecule. Such a substance is expected to elicit the desired antihypertensive effect due to a balanced relationship between β-adrenoceptor blockade and diuresis.

In order to achieve this objective, researchers have started combining the chemical structures of β-adrenoceptor blockers with those of diuretics. Whereas with regard to the β-blockers, it appears reasonable to start only with the 1-alkylamino-3-aryloxy-2-propanol structures, the possibilities for the diuretics are much more diverse. The numerous different structures of established diuretics allow a wide variety of molecular combinations. A selection of new compound classes concerning this new branch of research, which is still in a very early stage of development, is presented in Fig. 21.

From the structure of the compound class *(84)*, it can be seen that a 1-amino-3-phenoxy-2-propanol *(82)* as a β-blocker component is combined with the diuretic amiloride *(83)* via an alkylene bridge to form a hybrid molecule. Only small structural changes have been performed in the two components. Compared to *(84)*, in the substance class *(85)* the alkylene bridge has been extended by a second alkylene bridge including a carbonamide group. The two compound series contain drugs

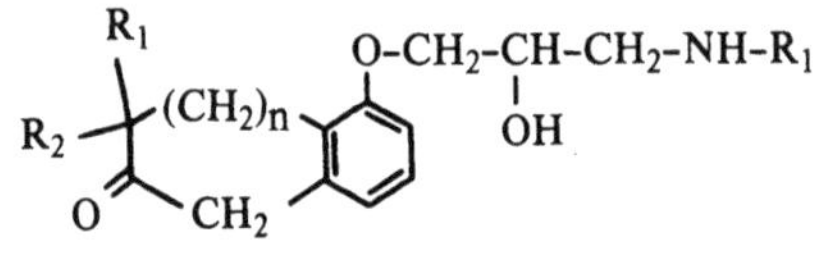

Sandoz 88)

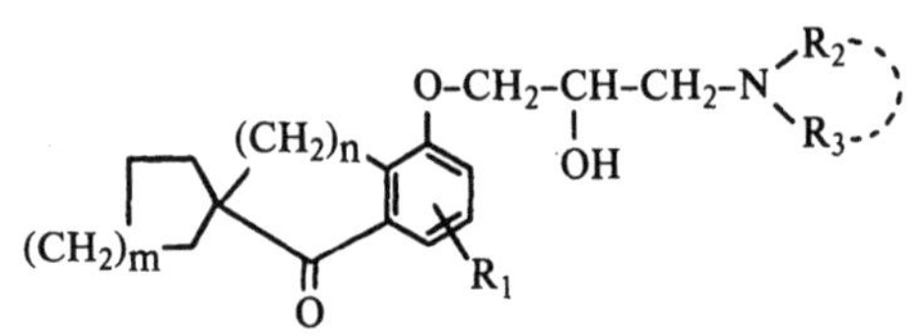

Sandoz 89)

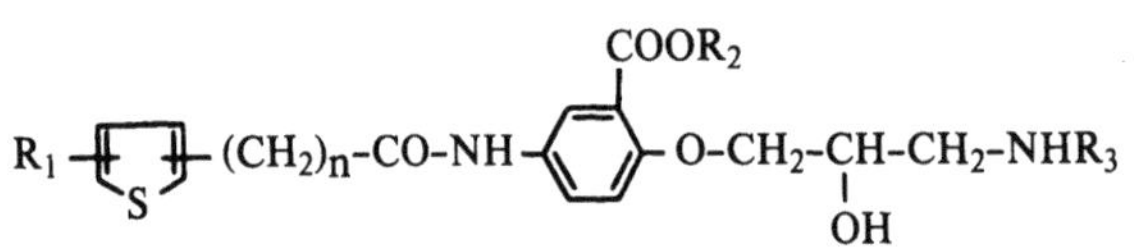

Carpibem 90)

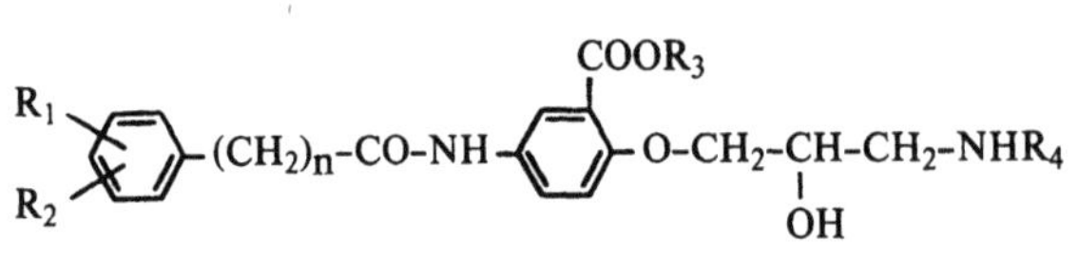

Carpibem 91)

Fig. 22. Beta adrenoceptor blocking drugs as antihypertensive agents with a diuretic effect

which elicit both β-adrenoceptor blockade and diuresis and at the same time exert an antihypertensive action[206, 207]. Formula *(87)* compounds have been synthesized in a quite different chemical way, but on the basis of the same considerations. The aromatic phenyl ring of the β-adrenoceptor blocker has been replaced here by the almost unchanged chlorothiazide skeleton *(86)*. The β-adrenoceptor blockade is combined with a blockade of the α-receptors, producing a pronounced antihypertensive effect. On local administration (glaucoma), this is also clearly shown by a marked reduction in intraocular pressure[208].

An activity spectrum consisting of β-blockade, diuresis and lowering of blood pressure has been described for substances from the class of 1-amino-3-phenoxy-2-propanol derivatives, which do not contain any diuretic structural moiety (Fig. 22).

For the two substance classes *(88)* and *(89)*, the carbocyclic annelation of the phenyl ring in the 2,3-position is typical. It contributes to an increase in lipophilicity which, on the other hand, is reduced by the respective carbonyl group. Besides antihypertensive and diuretic effects, metabolic activities were also registered[209, 210]. The compound classes *(90)* and *(91)* are typically substituted by an ester and a carbonamide function in the 2- and 4-positions of the phenyl ring. The activity profiles are similar. The β_1-selective substances are antihypertensives and diuretics with thiazide character[211, 212].

4.4 Enhancement of Antihypertensive Action by Structural Variations of β-Adrenoceptor Blocking Drugs with Vasodilatatory Properties

The antihypertensive effect of a β-adrenoceptor blocker can be intensified not only by means of a diuretic, but also by combining it with a vasodilator. The increase in heart rate elicited reflectorily by the vasodilator is appreciably attenuated or entirely abolished by a β-adrenoceptor blocker. Consequently, joint administration of the β-adrenoceptor blocker and a vasodilator is now also included in the phase plan of the WHO[178]. It has already been suggested that the properties of such a combination should be brought together in one drug molecule[213, 214]. Work is being pursued in numerous research centers. The available results are summarized below.

Highly active β-adrenoceptor blockers which may possibly already produce a vasoactive effect, e.g. bunitrolol *(92)*[87] served as leads for these molecular variations. Bunitrolol produces a distinct vasodilation in the hind legs of rats and dogs[215].

It can be seen from Fig. 23 that the bunitrolol molecule has been subjected to different structural variations. Above all, substitution of the tertiary butylamino group intensifies the vasodilatory effect.

In bucindolol *(93)*, a hydrogen atom of the tertiary butylamino residue of bunitrolol has been replaced by the indolyl residue[216]. With a pronounced β-blockade combined with simultaneous ISA, vasodilation[219] was observed after administration of bucindolol both in rats[217] and dogs[218]. A relaxant effect in the vascular smooth muscle has been described[220]. The S- and R-enantiomers bring about vasodilation, ISA and α- and β-blockade. However, the S-form is 250 times more active[221]. The chemical structure of the substance MK 761 *(94)* differs from bunitrolol in that the

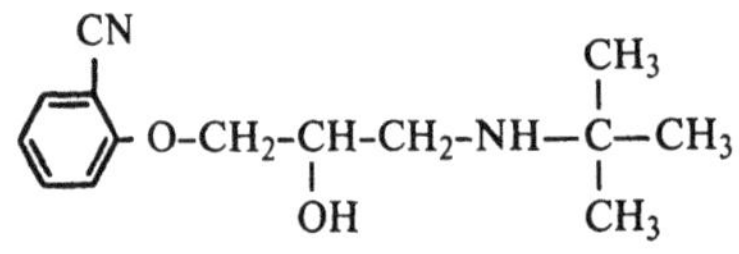

Bunitrolol
KOE 1366
Boehringer Ingelheim 92)

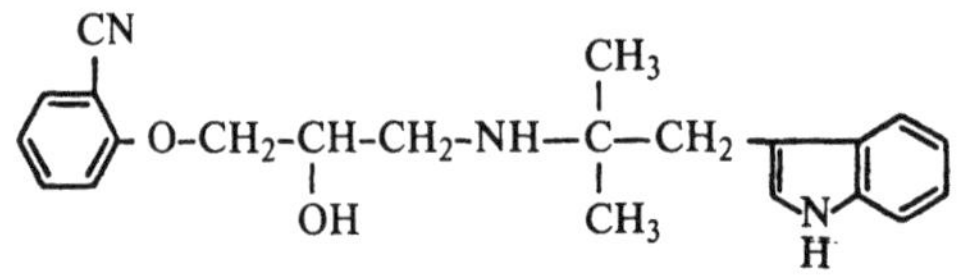

Bucindolol
MJ 13105-1
Mead Johnson/Bristol-Myers 93)

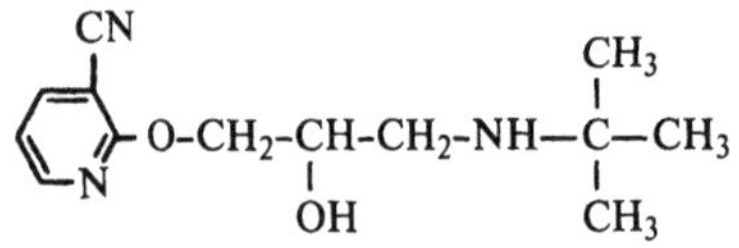

MK 761
Merck 94)

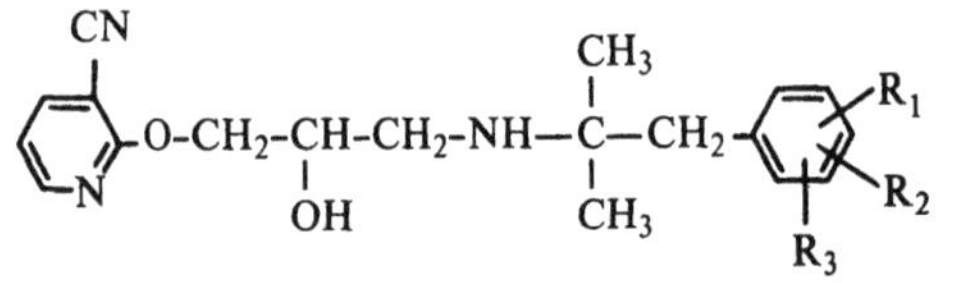

Bristol-Myers 95)

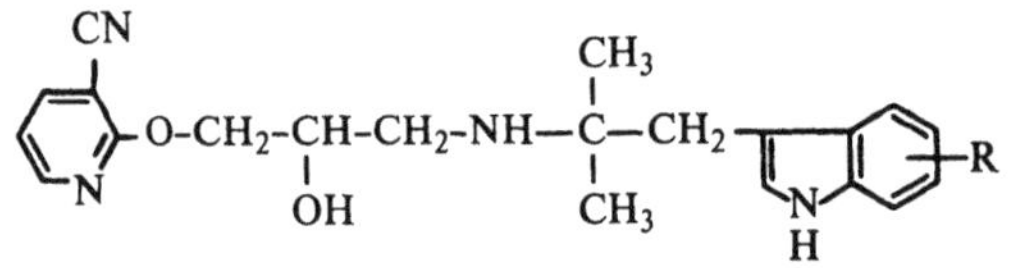

Hoechst 96)

Fig. 23. Beta adrenoceptor blocking drugs with vascular dilation; congeners of bunitrolol

phenyl group has been exchanged for the 2-pyridyl residue[222, 223]. A nonselective β-adrenoceptor blockade combined with ISA has been seen in animal studies[224]. In SH rats, the substance brings about a lowering of blood pressure without an effect on heart rate[225, 226]. The effect on blood pressure is based on vasodilation and β-blockade. The development was discontinued, since MK 761 produces teratogenic effects in rabbits[227]. The exchange of the phenyl residue in bucindolol for the pyridyl residue has resulted in compounds with the formula *(95)*. They are both isoproterenol antagonists and vasodilators[228]. Substitution of the tertiary butylamino group by a phenyl residue in compound MK 761 yields the substance class *(96)*. These pyridine derivatives are β-adrenoceptor blockers with a marked vasodilatory action and elicit pronounced lowering of blood pressure due to the double effect[228a]. Some prominent substances have been synthesized as R- and S-enantiomers.

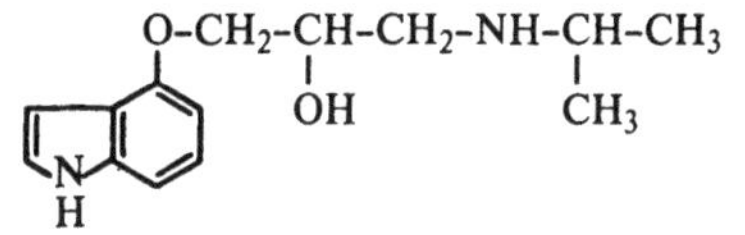

Pindolol
Sandoz
97)

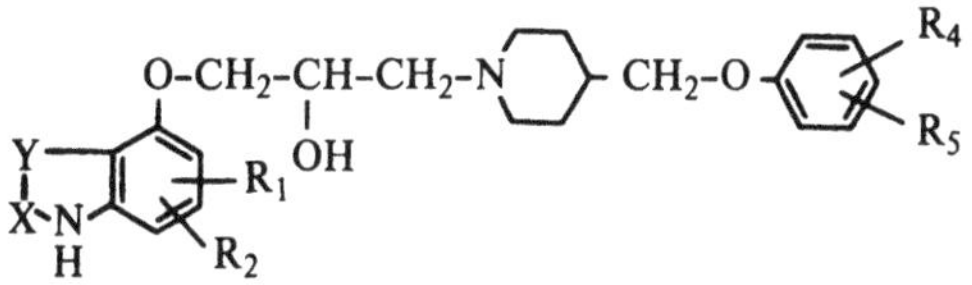

Boehringer Mannheim
98)

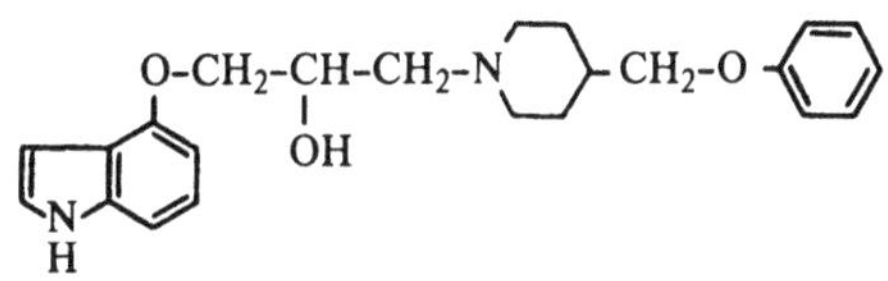

Mindolilol
BM 12434
Boehringer Mannheim
99)

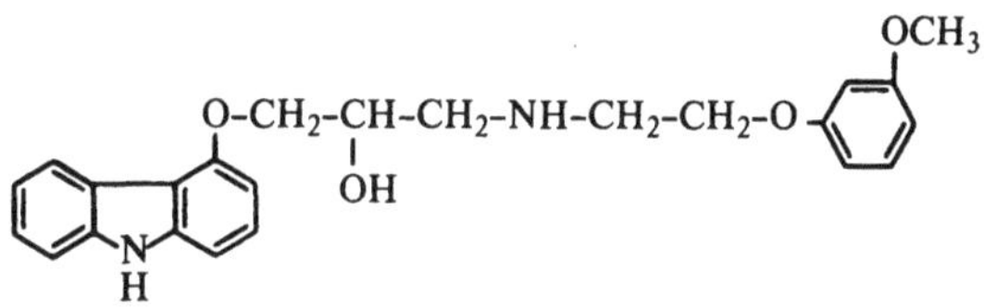

Carvedilol
BM 14190
Boehringer Mannheim
100)

Fig. 24. Vasoactive beta adrenoceptor blocking drugs; congeners of pindolol

Starting from the structure of pindolol, substance groups have been synthesized in order to obtain new vasoactive β-adrenoceptor blockers. Figure 24 shows selected examples.

Pindolol *(97)* is a potent β-adrenoceptor blocker[229, 230] which, for years, has been proving its efficacy in the treatment of angina pectoris and hypertension[231, 232]. By exchanging the isopropylamino group for a substituted piperidino residue, formula *(98)* compounds were synthesized[233]. Among these, substance BM 12434 *(99)* is being developed because of its activity profile. It reduces the peripheral vascular resistance with therapeutic β-blocker doses and brings about an increase in arterial blood flow and venous vascular capacity[234]. Due to venous dilation, the preload of the heart is reduced[235]. In carvedilol *(100)*, the indolyl residue of pindolol is replaced by the carbazolyl group, and the alkylamino group contains a phenolic ether[236]. This not only raises lipophilicity but also β_1-selectivity. In SH rats, the substance brings about a β-blockade and vasodilation[237]. In clinical investigations, a distinct lowering of blood pressure with slight orthostatic effects was observed[238].

4.5 Enhancement of Antihypertensive Action by Nitroester Substitution

It has been known for a long time that nitric acid esters ("nitrates") cause a vasodilation by acting on the vascular musculature. In consequence, there is an increase in blood volume on the venous side which reduces the reflux to the heart, and thus lowers the filling volume. The reduction of the diastolic wall tension associated with this, as well as the reduction of aortic pressure by dilation of the arteries, reduction of systolic wall tension and peripheral vascular resistance lead to a diminution of cardiac work. The oxygen requirement of the heart falls. Nitric acid esters are thus indicated in the treatment of angina pectoris attacks. They are also administered simultaneously with a β-adrenoceptor blocker or in a fixed combination with such a drug. Research work has also attempted here to combine the activity profile of two drugs in one agent[239)]. Figure 25 shows first results of work in this field which is still in progress.

Whereas in formulae *(101)* and *(102)* the nitric acid ester group ("nitrate") is linked with the 1-alkylamino-3-phenoxy-2-propanol via an alkyleneamide[240)] or an alkylene bridge[241)], the ester function in nipradilol (K-351) *(104)* is directly connected with the benzopyranyl ring. This substance under development was selected from compound class *(103)*[242)], because it causes a marked and long-lasting reduction in blood pressure in animal experiments[243)]. K-351 chiefly inhibits α-adrenoceptors[244)], whereby vasodilation was observed[245)]. At the same time, this substance brings about a pronounced β-adrenoceptor blockade and a nitrate-like vascular relaxation[239)].

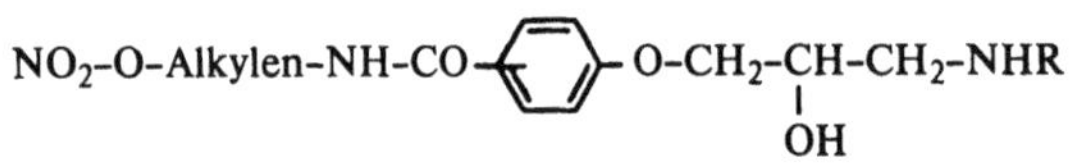

KOWA 101)

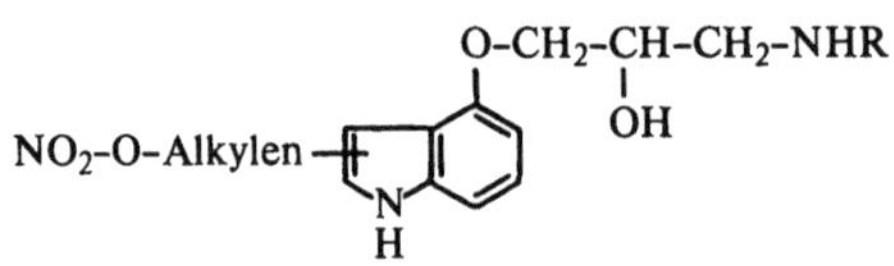

KOWA 102)

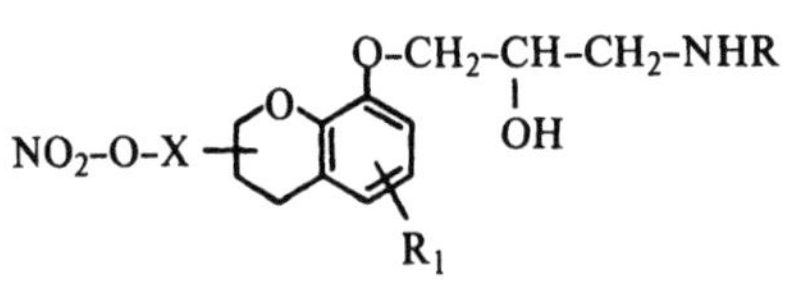

KOWA 103)

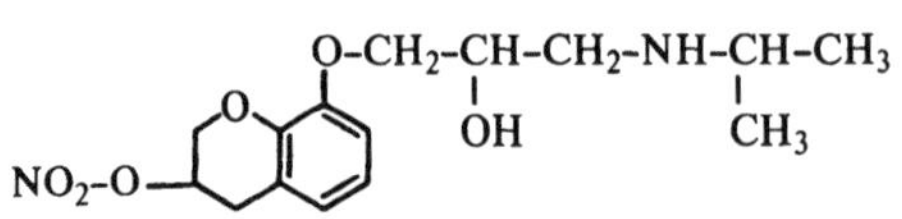

Nipradilol 104)
K-351
KOWA

Fig. 25. Beta adrenoceptor blocking drugs with vasodilation; derivatives of nitric acid esters

4.6 *Enhancement of Antihypertensive Action by Combination of the Structures of β-Adrenoceptor Blocking Drugs and a Vasodilator*

Combination of a β-adrenoceptor blocker with a vasodilator is planned in the WHO phase scheme for the treatment of hypertension. The improved activity spectrum achieved by the simultaneous administration of two drugs occasioned efforts to combine this profile in one molecule. Well-established vasodilators were included in the narrowed-down selection of possible partners for synthesis. Without major structural alterations, researchers were able to link the partners chemically to β-adrenoceptor blockers to form a "hybrid molecule". Combination with a β-adrenoceptor blocker is possible for example with hydrazine derivatives of the hydralazine or cadralazine type. Figure 26 shows the structures of the two vasodilators as well as of the new combination molecules.

The vasodilator hydralazine *(105)*[246, 247] served as the initial basis for the synthesis of compounds of the formula *(107)*[249]. In analogy to cadralazine *(106)*[248], hydrazinopyridazines could be converted into substances of formula *(108)*[250]. Both substance classes contain β-adrenoceptor blocker properties whose hypotensive effect is accompanied by vasodilation[249, 250]. The phenylpyridazine derivatives *(109)* and *(110)* are structural isomers with a different arrangement of the phenolether oxygen. In both substance classes, not only the structures of the vasodilator and the β-blocker are combined, but their activity profile is likewise characterized by β-blockade, vasodilation and the associated lowering of blood pressure[251, 252]. As a substance under development, SKF 92657 (prizidilol *(111)*) resulted from substance class *(110)*. After administration of the drug in man, the characteristics of β-blockade with simultaneous vasodilation and lowering of blood pressure were ascertained[253, 254]. The reduction in peripheral resistance to flow is associated with a lowering of heart rate and an increase in the stroke volume[255]. Hydrazinopyridazine derivatives which are linked to the β-blocker structure by an alkylene bridge were simultaneously synthesized in two research centers[256, 257]. They differ in the arrangement and nature of their substituents. In contrast to compounds of formula *(112)*, in substance class *(113)* the heterocycle is directly linked to the phenoxy residue of the β-blocker via an alkylene residue. In both compound series, the pronounced lowering of blood pressure is caused by β-blockade with vasodilation[258]. As in the substance classes *(108)* to *(110)*, the latter is attributable above all to the characteristic hydrazinopyridazine moiety.

4.7 *β-Adrenoceptor Blocking Drugs for Treatment of Glaucoma*

β-adrenoceptor blockers lower the intraocular pressure in glaucoma. This effect was already observed in 1967 after administration of propranolol[259]. The same effect could also be obtained later with newer preparations, e.g. with metoprolol[260]. Timolol[261] has been used very extensively in this field in the USA. Although most β-blockers are likely to be suitable for this indication, those agents without local

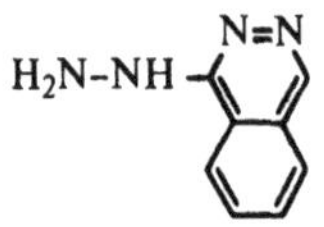

105)

Hydralazine
Ciba-Geigy

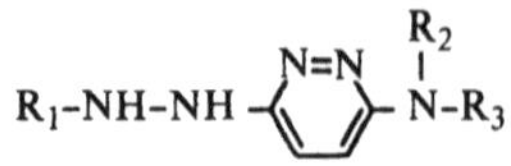

106)

Cadralazine-Typ
ISF

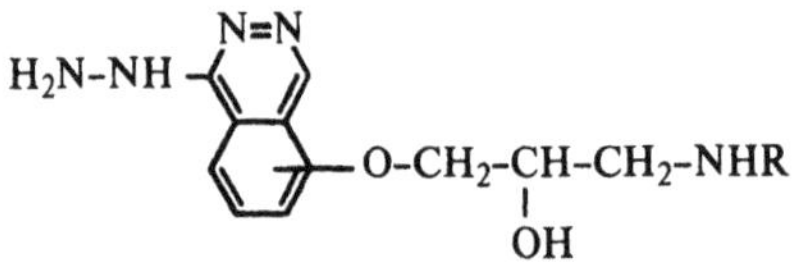

SKF 107)

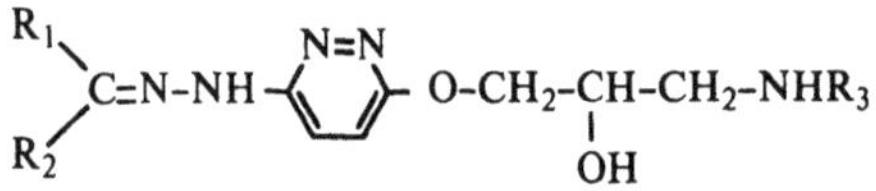

ISF 108)

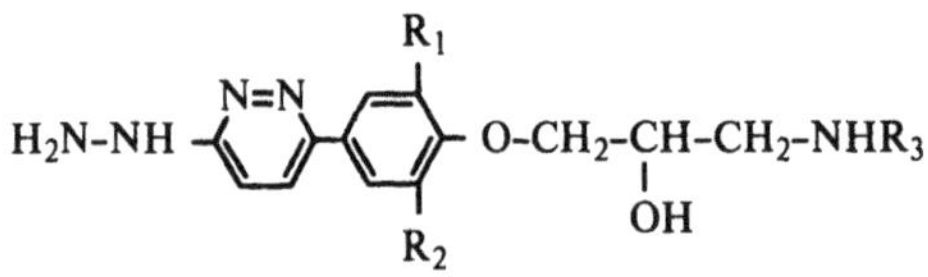

SKF 109)

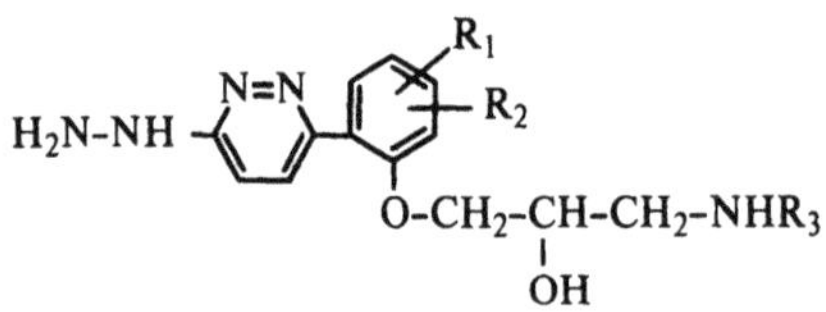

SKF 110)

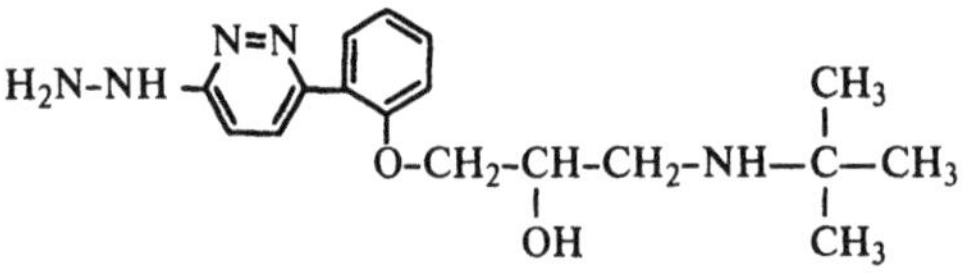

SKF 92657
Prizidilol
SKF 111)

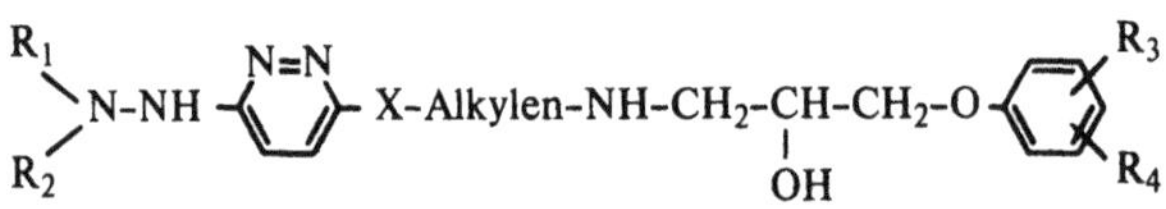

ICI/Teikoku 112)

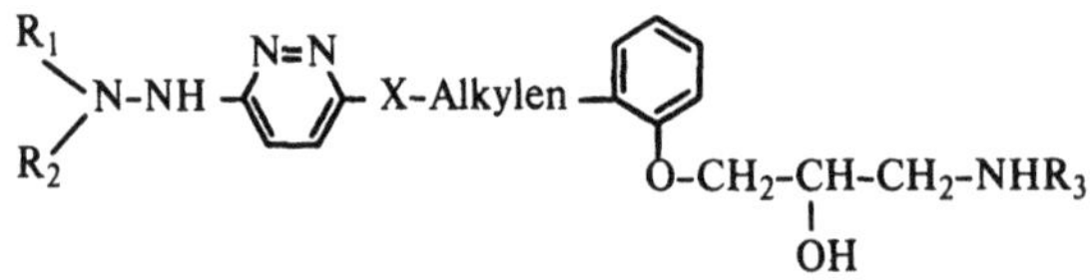

Teikoku 113)

Fig. 26. Beta adrenoceptor blocking drugs with vasodilation: hybridization of a beta blocking drug and a vasodilator

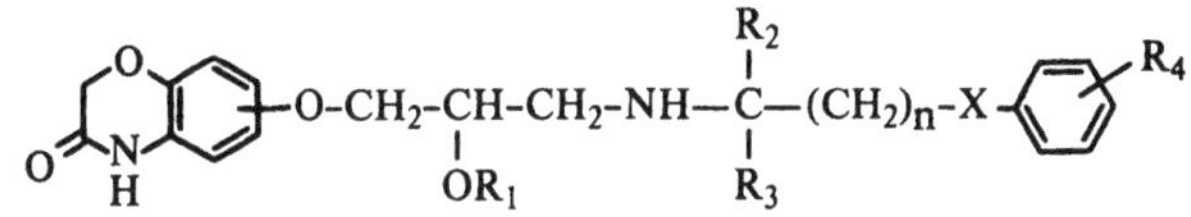

Merck 114)

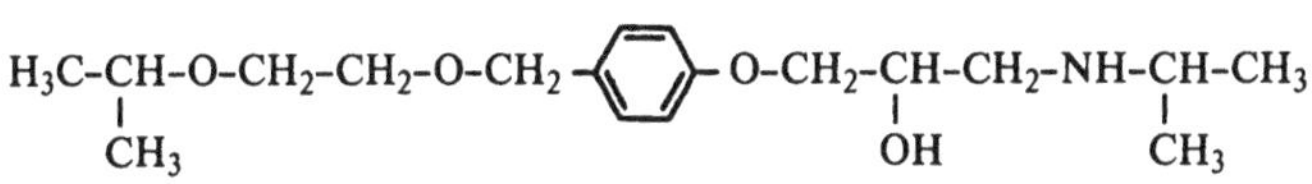

EMD 33512
Bisoprolol 115)
E. Merck

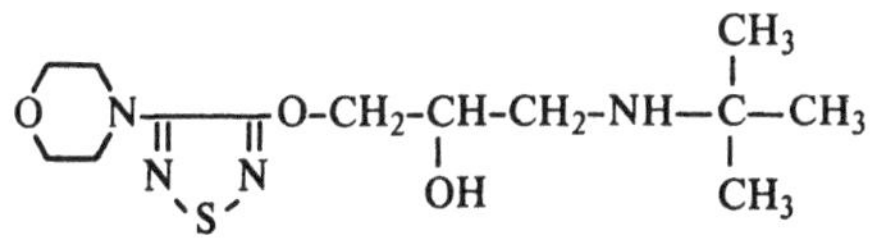

Timolol 116)
Merck

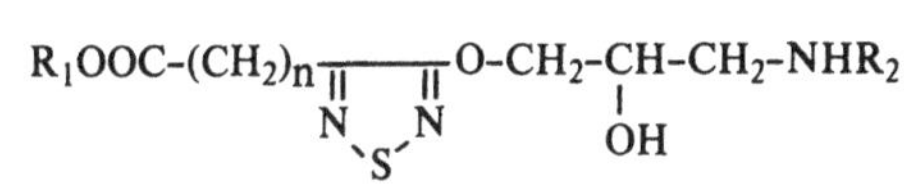

Cyanamid 117)

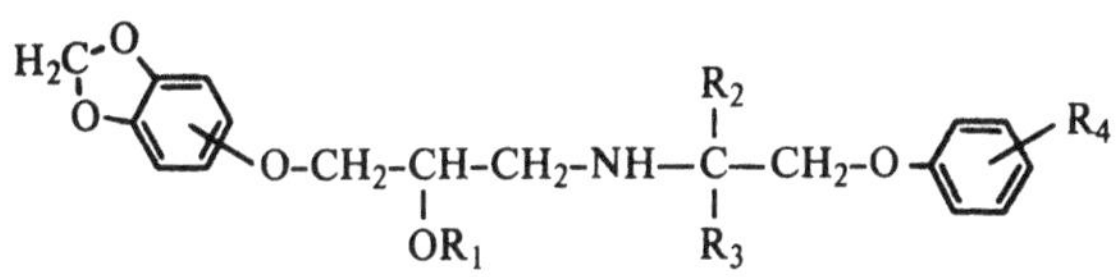

Merck 118)

Fig. 27. Development of beta adrenoceptor blocking drugs for the treatment of glaucoma

anesthetic effects are preferable. Following topical administration, the long-lasting reduction in intraocular pressure is hardly impaired by side effects. There has been no lack of attempts to develop special β-blockers for this area of application. A small selection of new substances is listed in Fig. 27.

Topical administration of the benzoxazinone derivative *(114)* leads to a reduction in intraocular pressure in glaucoma even at very low doses[262]. The selective β_1-adrenoceptor blocker bisoprolol *(115)* likewise reduces intraocular pressure without obvious side effects[263–265]. The chemical similarity of the thiadiazol derivatives *(117)*[268] to timolol *(116)*[266, 267] is evident. In analogy to timolol, it is planned to use them for local administration in the treatment of glaucoma[267]. Their side effects are slight. A reduction of the intraocular pressure is observed after administration of compounds of formula *(118)*, so that they are also intended for the treatment of glaucoma[269].

5 Summary

The high therapeutic value of the β-adrenoceptor blockers is chiefly attributable to their reliable effect in coronary heart disease, certain arrhythmias and hypertension. Due to the low incidence of side effects, they are suitable for long-term treatment in various indications. The discovery of specific properties in drug molecules of different structure (intrinsic sympathomimetic activity, β_1-adrenoceptor selectivity, membrane-stabilizing action) has stimulated chemical research into intensifying or eliminating such inherent qualities of the substances. On the one hand, the research was aimed at prolonging the duration of action and on the other hand developing short-acting β-adrenoceptor blockers. By structural variation, alteration of the lipophilic properties, different space filling of specific molecular regions and by isolation of optical enantiomers, attempts were made to reduce still further the incidence of side effects. Much of the research work was focused on β-adrenoceptor blocking substances which, as antihypertensives, bring about an increased and prolonged reduction in blood pressure. This objective was approached from various directions. The α/β-blockade of labetalol was the reason for improving the not completely satisfactory activity profile of this preparation. Further studies are necessary to achieve this objective. The combination of a β-adrenoceptor blocker with a diuretic recommended by the WHO as well as the German League for Control of Hypertension was taken as the basis for syntheses aimed at obtaining agents which combine the positive features of both drugs. There has been a parallel development in the combination of drug groups whose vasodilatory effect further intensifies the already present hypotensive effect of the β-blocker. Molecular variation alone has led to β-adrenoceptor blockers in which the hypotensive effect is increased by additional vasodilation.

The mechanisms by which β-adrenoceptor blockers reduce blood pressure have not yet been clarified. They are of a complex nature, and the reduction in cardiac output is likely to play a role. Moreover, a reduction of plasma renin in patients whose plasma renin levels show raised values is under discussion.

6 Outlook

25 years after the synthesis in summer 1960 of the first 1-alkylamino-3-phenoxy-2-propanols of use in cardiac therapy, research activity in the field of β-adrenoceptor blocking drugs has now reached its peak. They have become a beneficial and indispensable aid to treatment for cardiac patients.

For years now, they have also been proving their efficacy in the treatment of hypertension and are now recommended as the drug of first choice in this indication. They contribute to limiting the consequences of this disease of modern civilization. As antihypertensives, they reduce the mortality due to hypertension and the incidence of coronary heart disease.

Further problems in the treatment of severe and in some cases therapy-resistant hypertension still await solution. Research is working towards these objectives.

7 Acknowledgements

For many years of productive cooperation, my warmest gratitude goes to Dr. H. Stähle and Dr. W. Kummer. For long and expert technical assistance, I wish to thank Mr M. Doch, Mr L. Kundel, Mr H. Osterod and Mr M. Thumerer. I am very grateful to Dr. A. Engelhardt, Dr. H. Merz, Dr. D. Middleton, Dr. W. Taylor and Dr. W. Traunecker for meticulous perusal of the manuscript, for suggestions and stimulating discussions. Furthermore, I am indebted to Mrs B. Bailey, Mr R. Ordish and Mr D. J. Williams for their assistance with the translation.

This contribution is dedicated to the Liebrecht, Boehringer and von Baumbach families, proprietors of the Boehringer Ingelheim company on the occasion of its centenary in 1985.

8 References

1. Shanks, R. G., in: Discoveries in Pharmacology, Vol. 2, p. 37 (M. J. Parnham, Bruinvels, eds.). Amsterdam: Elsevier 1984
2. Morganroth, J., in: The Evaluation of Beta Blocker and Calcium Antagonist Drugs, p. 2 (J. Morganroth, E. N. Moore, eds.). The Hague–Boston–London: Nijhoff 1982
2a. Abrams, W. B., in: The Evaluation of beta Blocker and Calcium Antagonist Drugs, p. 7 (J. Morganroth, E. N. Moore, eds.). The Hague–Boston–London: Nijhoff 1982
3. Lohmann, F. W. (ed.), in: Die klinische Bedeutung der β_1-Selektivität, p. 9. München: Verlag für angewandte Wissenschaften 1984
4. Kewitz, H., in: β-Rezeptorenblockade, p. 100 (F. W. Lohmann, ed.). Berlin–Heidelberg–New York: Springer 1982
5. Bolte, H.-D. (ed.), in: Therapie mit Beta-Rezeptorenblockern, p. 19. Berlin–Heidelberg–New York: Springer 1979
6. Lüderitz, B., in: β-Rezeptorenblockade, p. 76 (F. W. Lohmann, ed.). Berlin–Heidelberg–New York: Springer 1982
7. Prichard, B. N. C.: Brit. J. Clin. Pharmacol. *13,* 51 (1982)
8. Prichard, B. N. C., in: Alpha and Beta Adrenoceptors and the Cardiovascular System, p. 178 (W. Kobinger, R. P. Ahlquist, eds.). Princeton: Excerpta Medica 1984
9. Simon, H.: Arzneim.-Forsch. *33,* 277 (1983)
10. Lydtin, H., Lohmöller, G., in: Beta Blockers, p. 138. Bern–Stuttgart–Wien: Huber 1980
11. Lands, A. M., et al.: Nature *214,* 597 (1967)
12. Dunlop, D., Shanks, R. G.: Brit. J. Pharmacol. *32,* 201 (1968)
13. Åblad, B., et al., in: Die klinische Bedeutung der β_1-Selektivität, p. 13 (Lohmann, F. W., ed.). München: Verlag für angewandte Wissenschaften 1984
14. Wolff, H. P., Schrey, A. (eds.), in: Beta-Blocker, Fragen und Antworten, Expertengespräch Triberg 1982, p. 4. München–Wien–Baltimore: Urban & Schwarzenberg 1984
15. Mutschler, E. (ed.), in: Arzneimittelwirkungen, p. 262. Stuttgart: Wissenschaftliche Verlagsgesellschaft mbH 1981
16. Gibson, D. G.: Drugs *7,* 8 (1974)
17. Schier, O., et al.: Progr. Drug Res. *25,* 9 (1981)
18. Evans, D. B., et al.: Ann. Rep. Med. Chem. *14,* 81 (1979)
19. Ahlquist, R. P.: Progr. Drug Res. *20,* 27 (1976)
20. Weetman, D. F.: Drugs Today *13,* 261 (1977)
21. Kobinger, W., Ahlquist, R. P. (eds): Alpha and Beta Adrenoceptors and the Cardiovascular System. Princeton: Excerpta Medica 1984

22. Borchard, U. (ed.): Pharmakologie der β-Rezeptorenblocker. Basel: Aesopus 1983
23. Ahlquist, R. P.: Am. J. Physiol. *153,* 586 (1948)
24. Shanks, R. G., in: Discoveries in Pharmacology, Vol. 2, p. 41 (M. J. Parnham, J. Bruinvels, eds.). Amsterdam: Elsevier 1984
25. Mills, J.: US 2938921 (1960)
26. Powell, C. E. Slater, I. H.: J. Pharmacol. Exp. Ther. *122,* 470 (1958)
27. Moran, H. C., Perkins, M. E.: J. Pharmacol. Exp. Ther. *124,* 223 (1958)
28. Shanks, R. G., in: Discoveries in Pharmacology, Vol. 2, p. 68 (M. J. Parnham, J. Bruinvels, eds.). Amsterdam: Elsevier 1984
29. Scheuing, G., Thomä, O.: DRP 723278 (1939)
30. Köppe, H.: Postgrad. Med. J. Suppl. (1) *53,* 22 (1977)
31. Engelhardt, A.: Unpubl. results 1961
32. Köppe, H., et al.: Belg. P. *652,* 336 (1964)
33. Engelhardt, A.: Naunyn Schmiedeberg's Arch. exp. Pathol. Pharmacol. *250,* 245 (1965)
34. Shanks, R. G., et al.: Nature *212,* 88 (1966)
35. Nayler, W. G., et al.: J. Pharmacol. Exp. Ther. *165,* 22 (1969)
36. Somani, P., et al.: Eur. J. Pharmacol. *14,* 209 (1971)
37. Wirtzfeld, A., et al.: Z. Kardiol. *71,* 619 (1982)
38. Köppe, H., et al.: JP 796550 (1964)
39. Engelhardt, A.: Unpublished results 1963
40. Carlsson, E., et al.: Arch. Pharmacol. *300,* 101 (1977)
41. Dahlstrom, V., et al.: Eur. J. Clin. Pharmacol. *24,* 495 (1983)
42. Graffner, C., et al.: Eur. J. Pharmacol. *20,* 91 (1981)
43. Ostermayer, F., et al.: EP 15505 (1980)
44. Ostermayer, F., et al.: EP 45964 (1982)
45. Fuhrer, W., et al.: J. Med. Chem. 27, 831 (1984)
46. Carlsson, E., et al.: EP 64487 (1982)
47. Barlow, J. J., et al.: J. Med. Chem. *23,* 624 (1980)
48. Drugs in the Future *7,* 571 (1982); *8,* 736 (1983)
49. Barlow, J. J., et al.: J. Med. Chem. *24,* 315 (1981)
50. Louis, W. J.: EP 105838 (1984)
51. Poittevin, A., et al.: DOS 2514630 (1975)
52. Lövgren, K., et al.: J. Med. Chem. *23,* 624 (1980)
53. Dunlop, D., Shanks, R. G.: Brit. J. Pharmacol. *32,* 201 (1968)
54. Lands, A. M., et al.: Nature *214,* 597 (1967)
55. Carlsson, E., et al.: Life Sci. II Part 1, 953 (1972)
56. Waal Manning, H. J., et al.: Lancet, 1971 – II 1264
57. Mutschler, E.: Dtsch. Apothekerz. *123,* 1723 (1983)
58. Smith, U., in: Die klinische Bedeutung der β_1-Selektivität, p. 55 (Lohmann, F. W., ed.). München: Verlag für angewandte Wissenschaften 1984
59. Holm, G. S., et al.: Acta Med. Scand. *208,* 187 (1980)
60. Howe, R., et al.: DOS 1643425 (1968)
61. Smith, L. H.: DOS 1936693 (1970)
62. Crowther, et al.: J. Med. Chem. *14,* 511 (1971)
63. Lydtin, H., in: Beta Blockers, p. 149 (Lydtin, H., Lohmöller, G., eds.). Bern–Stuttgart–Wien: Huber 1980
64. Barrett, A. M., et al.: DE 2007751 (1971)
65. Graham, B. R., et al.: Brit. J. Pharmacol. *49,* 154 (1973)
66. Marmo, E.: Drugs exp. clin. Res. *6,* 639 (1980)
67. Brunner, H., et al.: Klin. Wschr. *56,* Suppl. 1, 107 (1978)
68. Isbary, J., et al.: Z. Kardiol. *67,* 857 (1978)
69. Myers, M. G., et al.: Clin. Pharmacol. Ther. *27,* 756 (1980)
70. Wooldridge, K., et al.: DOS 1815808 (1970)
71. Basil, B., et al.: J. Med. Chem. *19,* 399 (1976)
72. Watson, R. D. S., et al.: Brit. J. Pharmacol. *7,* 557 (1979)
73. Main, B. G.: J. Chem. Techn. Biotechnol. *32,* 617 (1982)
74. Holmes, A., et al.: DOS 2259489 (1971)

75. Hoefle, M. L., et al.: J. Med. Chem. *18*, 148 (1975)
76. Gross, G. J., et al.: J. Cardiovasc. Pharmacol. *1*, 139 (1979)
77. Augstein, J., et al.: DOS 1957706 (1970)
78. Augstein, J., et al.: J. Med. Chem. *16*, 1245 (1973)
79. Pfisterer, M., et al.: Schweiz. Med. Wschr. *106*, 549 (1976)
80. Rzeszotarski, W. J., et al.: J. Med. Chem. *26*, 644 (1983)
81. Rzeszotarski, W. J., et al.: J. Med. Chem. *22*, 735 (1979)
82. Cohnen, E.: DE 3030530 (1982)
83. Cohnen, E.: DE 3030573 (1982)
84. Bercher, H., et al.: DD 153682 (1982)
85. Zoelss, G.: DOS 2458624 (1974)
86. Zoelss, G.: Arzneim.-Forsch. Drug Res. *33*, 2 (1983)
87. Köppe, H., et al.: DAS 1593782 (1972)
88. Köppe, H., et al.: DAS 1643262 (1972)
89. Köppe, H., et al.: DOS 2309887 (1974)
90. Köppe, H., et al.: DAS 1543391 (1972)
91. Muacevic, G., Traunecker, W.: Publication in progress
91a. Mylecharane, E. J., et al.: Eur. J. Pharmacol. *21*, 375 (1973)
92. Köppe, H., et al.: EP 101498 (1981)
93. Köppe, H., et al.: EP 101499 (1981)
94. Köppe, H., et al.: EP 102294 (1981)
95. Köppe, H., et al.: EP 107519 (1982)
96. Köppe, H., et al.: EP 107536 (1982)
97. Traunecker, W.: Arzneim.-Forsch. Drug Res. *35* (I) 1a, 376 (1985)
98. Brogden, R. C., et al.: Drugs *14*, 321 (1977)
99. Clausen, N., et al.: Brit. J. Clin. Pharmacol. *7*, 379 (1979)
100. Cohnen, E., et al.: DE 3151201 (1983)
101. Manoury, P. M., et al.: DOS 2649605 (1980)
102. Bianchetti, G., et al.: Arzneim.-Forsch. *30*, 1912 (1980)
103. Balnave, K., et al.: Brit. J. Clin. Pharmacol. *11*, 171 (1981)
104. Assael, B. M., et al.: Clin. Pharmacol. Ther. *29*, 232 (1981)
105. Pouleur, H., et al.: J. Cardiovasc. Pharmacol. *6*, 720 (1984)
106. Regnier, F., et al.: Therapie *38*, 155 (1983)
107. Berthold, R., et al.: EP 52072 (1982)
108. Gustafsson, et al.: EP 85286 (1983)
109. Machin, P. J., et al.: EP 41233 (1981)
110. Machin, P. J., et al.: J. Med. Chem. *26*, 1570 (1983)
111. Manoury, P., et al.: EP 61379 (1982)
112. Muro, T., et al.: WO 8401945 (1984)
113. Machin, P. J.: EP 34831 (1981)
114. Machin, P. J., et al.: J. Med. Chem. *27*, 503 (1984)
115. Lindner, E.: Arzneim.-Forsch. *34*, 270 (1984)
116. Fritsch, W., et al.: DE 2623314 (1977)
117. Witte, P. U., et al.: Naunyn Schmiedeberg's Arch. Pharmacol. *322* (Suppl.) Abst. 522 (1983)
118. Tominaga, M., et al.: DOS 2711719 (1977)
119. Tominaga, M.: Chem. Pharm. Bull. *29*, 2166 (1981)
120. Sugawara, K., et al.: Arch. int. Pharmacodyn. *240*, 294 (1979)
121. Yo, E., et al.: Chem. Pharm. Bull. *29*, 2157 (1981)
122. Baldwin, J. J., et al.: J. Med. Chem. *26*, 950 (1983)
123. Baldwin, J. J., et al.: VIIIth Int. Symp. Med. Chem. Uppsala, Abstracts 11 (1984)
124. Ostermayer, F., et al.: EP 45964 (1982)
125. Fuhrer, W., et al.: DE 3314196 (1983)
126. Jaeggi, K. A., et al.: GB 2014146 (1979)
127. Lehmann, D., et al.: DE 2844497 (1978)
128. Casagrande, C., et al.: EP 116825 (1984)
129. Mennini, T., et al.: J. Mol. Cell. Cardiol. *16*, Suppl. (1984)
130. Nicola, M., et al.: VIIIth Int. Symp. Med. Chem. Uppsala, Abstracts 23 (1984)

131. Smith, L. H.: DE 2362568 (1974)
132. Smith, L. H.: DE 2525133 (1975)
133. Smith, H. J., et al.: J. Pharm. Exp. Ther. *226,* 211 (1983)
134. Norris, S. C., et al.: Brit. J. Clin. Pharmacol. *18,* 271 (1984)
135. Harry, J. D., et al.: Brit. J. Clin. Pharmacol. *18,* 291 (1984)
136. Large, M. S., et al.: J. Med. Chem. 25, 1286 (1982)
137. Majid, P. A., et al.: J. Cardiovasc. Pharmacol. *2,* 435 (1980)
138. Matier, W. L., et al.: DE 3217206 (1982)
139. Raabe, T., et al.: DE 3048487 (1982)
140. Raabe, T., et al.: DOS 2819629 (1978)
141. Suntory Ltd. J 59164786 (1984)
142. Howe, R., Shanks, R. G.: Nature *210,* 1336 (1966)
143. Schmitt, K., et al.: DOS 1668055 (1971)
144. Härtfelder, G., et al.: Arzneim.-Forsch. *22,* 930 (1972)
145. Vierhapper, H., et al.: Arzneim.-Forsch. *30,* 1008 (1980)
146. von Leitner, E. R., et al.: Eur. J. Cardiol. *12,* 121 (1980)
147. Shavel, J., et al.: DOS 1948144 (1969)
148. Shavel, J., et al.: DOS 2046043 (1971)
149. Schwender, C. F., et al.: J. Med. Chem. *13,* 684 (1970)
150. Commarato, M. A., et al.: Arch. Int. Pharmacodyn. Ther. *226,* 205 (1977)
151. Duzman, E., et al.: Am. J. Ophthalmol. *94,* 318 (1982)
152. Quast, U., et al.: Arzneim.-Forsch. *34,* 579 (1984)
153. Bartsch, W.: DE 3300933 (1984)
154. Leinert, H.: DE 3319027 (1984)
155. Innis, R. B., et al.: Life Sci. *24,* 2255 (1979)
156. Bartsch, W., et al.: Arzneim.-Forsch. *27,* 1022 (1977)
157. Ghirardi, P., et al.: Cardiovasc. Pharmacol. *2,* 471 (1980)
158. Grosso, F., et al.: Acta Ther. *6,* 87 (1980)
159. Nicaise, P., et al.: Curr. Ther. Res. *30,* 93 (1981)
160. Ehrhardt, P. W., et al.: J. Med. Chem. *25,* 1402 (1982)
161. Badia, A.: Drugs of the Future *9,* 183 (1984)
162. Sakanashi, M.: Trends Pharmacol. Sci. *234* (1981)
163. Maseri, A., et al.: N. Engl. J. Med. *299,* 1271 (1978)
164. Kam, S., et al.: J. Med. Chem. *27,* 1007 (1984)
165. Zaroslinski, J., et al.: Life Sci. *31,* 899 (1982)
166. Ehrhardt, P. W., et al.: US 4450173 (1982)
167. Ehrhardt, P. W., et al.: EP 53435 (1982)
168. Matier, W. L.: WO 83/00043 (1981)
169. Kam, S., et al.: WO 83/01868 (1981)
170. Jacobi, A., et al.: J. Pharm. Sci. *72,* 710 (1983)
171. Ehrhardt, P. W., et al.: J. Med. Chem. *26,* 1109 (1983)
172. Black, J. W.: J. Pharm. Pharmacol. *12,* 87 (1960)
173. Prichard, B. N. C.: Brit. Med. J. *1,* 1227 (1964)
174. Prichard, B. N., et al.: Brit. Med. J. *2,* 725 (1964)
175. Hansson, L., in: β-Rezeptorenblockade, p. 78 (Lohmann, F. W., ed.). Berlin–Heidelberg–New York: Springer 1982
176. Prichard, B. N. C., et al.: Cardiology *66* (Suppl.), 1 (1980)
177. Wolff, H. P., in: Beta-Blocker, Fragen und Antworten, p. 22 (Wolff, H. P., Schrey, A., eds.). München–Wien–Baltimore: Urban & Schwarzenberg 1984
178. Gross, F.: Inn. Med. *9,* 427 (1982)
179. Mutschler, E.: Arzneimittelwirkungen, p. 262. Stuttgart: Wiss. Verlagsgesellschaft 1981
180. Diamond, J.: US 4434174 (1982), EP 34116 (1981)
181. Köppe, H., et al.: DOS 2644833 (1976)
182. Hoefke, W., et al.: Naunyn Schmiedeberg's Arch. Pharmacol. *325,* 54 (1984)
183. Wiedemann, I.: Arzneim.-Forsch. *33,* 861 (1983)
184. Palluk, R., et al.: Naunyn Schmiedeberg's Arch. Pharmacol. *329,* R 79, Abst. 314 (1985)
185. Werner, L. H., et al.: EP 96006 (1983)

186. Hamilton, R. W., et al.: DE 3122814 (1982)
187. Obase, H., et al.: Chem. Pharm. Bull. *30*, 474 (1982)
188. Clark, R. D., et al.: J. Med. Chem. *26*, 657 (1983)
189. Lunts, L. H., et al.: DOS 2616403 (1978)
190. Farmer, J. B., et al.: Brit. J. Pharmac. *45*, 660 (1972)
191. Richards, D. A., et al.: Brit. J. Clin. Pharm. 3, 967 (1976)
192. Sybertz, E. J., et al.: J. Pharmacol. Exp. Ther. *218*, 435 (1981)
193. Borchard, U.: Selecta *52*, 4313 (1984)
194. Brogden, R. N., et al.: Drugs *15*, 251 (1978)
195. Lund-Johansen, P., et al.: Brit. J. Clin. Pharmacol. *7*, 169 (1979)
196. Vlackakis, N. D., et al.: Pharmacology *22* (3), Abstract 176 (1980)
197. Schechter, P. J., et al.: J. Cardiovasc. Pharmacol. *5*, 143 (1983)
198. Cheng, H. C., et al.: Pharmacology *22* (3), Abstract 31 (1980)
199. Kubo, K., et al.: Arzneim.-Forsch. *33*, 926 (1983)
200. Takizawa, H., et al.: DE 3010752 (1980)
201. Berthold, R.: CH 634050 (1983)
202. Bhattacharjee, et al.: Brit. J. Pharmacol. *80*, 491 (1983) (Proc. Suppl.)
203. Berthold, R.: DE 3200304 (1982)
204. Takahashi, T., et al.: EP 113910 (1984)
205. Gross, F.: Inn. Med. *9*, 429 (1982)
206. Barlow, J. J., et al.: EP 86564 (1983)
207. Barlow, J. J., et al.: EP 57572 (1982)
208. Atkinson, J. G., et al.: US 4442094 (1984)
209. Payne, T. G., et al.: DOS 2810869 (1978)
210. Berthold, R., et al.: DOS 2802758 (1979)
211. Bouley, E., et al.: EP 99822 (1984)
212. Bouley, E., et al.: EP 122827 (1984)
213. Gross, F.: Brit. J. Clin. Pharmacol. *13*, 133 (1982)
214. Köppe, H.: Plenary lecture, 56th Chemical Conference and Exhibition, Med. Chem. Div., Montreal 1973
215. Traunecker, W., Häselbarth, V., Brückner, S., in: Pharmacology of Antihypertensive Drugs, p. 307 (A. Scriabine, ed.). New York: Raven Press 1980
216. Kreighbaum, W. E., et al.: DOS 2830884 (1979)
217. Kreighbaum, W. E., et al.: J. Med. Chem. *23*, 285 (1980)
218. Deitchman, D., et al.: Arch. Int. Pharmacodyn. *247*, 76 (1980)
219. Deitchman, D., et al.: Eur. J. Pharmacol. *61*, 263 (1980)
220. Oates, H. F., et al.: Arch. Int. Pharmacodyn. Ther. *251*, 95 (1981)
221. Dolak, T. M., et al.: VIIIth Int. Symp. Med. Chem. Uppsala 1984, Abstracts 17
222. Baldwin, J. J.: DOS 2556110 (1976)
223. Baldwin, J. J.: DOS 2726754 (1978)
224. Sweet, C. S., et al.: J. Pharmacol. Exp. Ther. *211*, 195 (1979)
225. Sweet, C. S., et al.: J. Pharmacol. Exp. Ther. *211*, 200 (1978)
226. Scriabine, A., et al.: Experientia *35*, 1634 (1979)
227. Baldwin, J. J., et al.: J. Med. Chem. *22*, 1284 (1979)
228. Kreighbaum, W.: DE 3331612 (1984)
228a. Knoll, J., et al.: DE 3301198 (1984)
229. Troxler, F., et al.: DOS 1620342 (1966)
230. Saameli, K.: Helv. Physiol. Acta *25*, 219 (1967)
231. Hill, R. C., et al.: Brit. J. Pharmacol. *36*, 368 (1969)
232. Wolff, H. P.: Dtsch. Med. Wschr. *100*, 1726 (1975)
233. Kampe, W., et al.: DOS 2651574 (1978), DOS 2737630 (1979), Circul. *45*, 571 (1972)
234. v. Möllendorff, E., et al.: Clin. Sci. *61*, 477 (1981)
235. v. Möllendorff, E., et al.: Münch. Med. Wschr. *123*, 764 (1981)
236. Wiedemann, F., et al.: DOS 2815926 (1979)
237. Sponer, G., et al.: Naunyn Schmiedeberg's Arch. Pharmacol. *32* (Suppl.), Abstr. 81 (1982).
238. Strein, K., et al.: Naunyn Schmiedeberg's Arch. Pharmacol. *325* (Suppl.), Abstr. 207 (1984)
239. Shiratsuchi, M., et al.: VIIIth Int. Symp. Med. Chem. Uppsala, Abstr. 25 (1984)

240. Shiratsuchi, M., et al.: EP 83256 (1983)
241. Shiratsuchi, M., et al.: EP 105763 (1984)
242. Shiratsuchi, M., et al.: EP 42299 (1981)
243. Shimitsu, S., et al.: Jap. J. Pharmacol. *32* (Suppl.), 135 (1982)
244. Asada, H., et al.: J. Pharmacol. Exp. Ther. *223,* 560 (1982)
245. Shirasawa, Y., et al.: Jap. J. Pharmacol. *32* (Suppl.), 175 (1982)
246. Druey, A., et al.: Helv. *34,* 195 (1951); Angew. Chem. *70,* 5 (1958)
247. Gross, F., et al.: Experientia *6,* 19 (1950)
248. Carpi, C., et al.: US 4002753 (1976)
249. Roe, A. M., et al.: DOS 2705414 (1977)
250. Dorigotti, L., et al.: EP 18016 (1980)
251. Coates, W. J., et al.: Brit. P 1548601 (1979)
252. Coates, W. J., et al.: DOS 2527066 (1976)
253. Larsson, R., et al.: Clin. Pharmacol. Ther. *29,* 588 (1981)
254. Luscher, T., et al.: Eur. J. Clin. Pharmacol. *23,* 411 (1982)
255. Lund-Johansen, P., et al.: J. Cardiovasc. Pharmacol. *4,* 1012 (1982)
256. Jasuda, K., et al.: EP 97202 (1984)
257. Briggs, M. T., et al.: EP 92945 (1983)
258. Teikoku Hormone Ltd.: JP 56142272 (1981); JP 56169675 (1981)
259. Phillips, C.I., et al.: Brit. J. Ophthalmol. *51,* 222 (1967)
260. Bietti, G. B., et al.: Klin. Mbl. Augenheilk. *170,* 824 (1977)
261. Zimmerman, T. J., et al.: Arch. Ophthalmol. *95,* 601 (1977)
262. Atkinson, J. G., et al.: US 4349673 (1982)
263. Harting, J.: EP 68160 (1982)
264. Tattensfield, A. E.: Brit. J. Pharmacol. *27,* 135 (1984)
265. Prager, G., et al.: Dtsch. Med. Wschr. *109,* 1914 (1984)
266. Wasson, B. K., et al.: J. Med. Chem. *15,* 651 (1972).
267. Heel, R. C., et al.: Drugs *17,* 38 (1979)
268. Matier, W. L., et al.: WO 8402525 (1984)
269. Baldwin, J. J., et al.: US 4472427 (1984)

NMR Analysis of Cancer Cells

Carolyn E. Mountford, Kerry T. Holmes and Ian C. P. Smith*

Ludwig Institute for Cancer Research (Sydney Branch), Blackburn Building, University of Sydney, Sydney, N.S.W. 2006, Australia

Nuclear magnetic resonance has been used to study cancer cells for almost two decades. Much confusion has been generated in the past due to heterogeneity of cell populations and to differences in sample preparation. Improved instrumentation and a better understanding of the complex growth patterns of cancer cells has led to more accurate measurements and to better control of samples. A brief overview of the NMR method is presented.

NMR spectroscopy may now be used to follow the cytoplasmic composition of cancer cells. A review of the utility of ^{1}H, ^{31}P and ^{19}F NMR spectroscopy in cancer research is given. The results from ^{31}P NMR are particularly promising, showing great sensitivity to the nature of a tumour and its response to therapy. NMR of ^{19}F is valuable in following the pharmacokinetics of anticancer drugs and in determining the products of their metabolism.

^{1}H NMR of cancer cells and tumours yields spectra with surprisingly narrow resonances, now known to be due to triglycerides bound to the plasma membrane in unique organelles. The relaxation behaviour of the methylene resonances in these spectra is shown to be an indicator of the metastatic potential of various cancer lines. The intensity of this resonance is a sensitive indicator of the degree of drug resistance of a cell line. The origins and interpretations of these spectra are described in detail, as are experimental protocols for sample preparation.

The potential for a better understanding of the biochemistry, biology, and medicine of cancer via the three major emphases of NMR – metabolism, membrane structure, and morphological imaging – is excellent.

* Division of Biological Sciences, National Research Council, Ottawa, K1A 0R6, Canada

Progress in Clinical Biochemistry
and Medicine, Vol. 3

A. Introduction

Nuclear magnetic resonance (NMR) is having a profound effect on medical diagnosis. It provides a safe way of distinguishing normal from pathological tissue and in some cases the molecular detail obtained also allows the cause of the tissue alteration to be determined. One exception to the success of NMR in medicine is in the field of Oncology. This is somewhat of a surprise since tumour detection was cited in the first publication on NMR imaging[1]. Although NMR imaging techniques are currently used to locate and diagnose tumours a justifiable concern exists over the wide range of relaxation parameters generated by imaging laboratories for normal, benign and malignant tissue. A statistical analysis by Bottomley et al.[2] shows that tissue relaxation times alone are inadequate for diagnosis of cancer but are of use as an indication of a pathological abnormality in the NMR image. Similarly the malignancy index proposed by Damadian is unsatisfactory because other non-cancerous pathological tissues have elevated T_1 or T_2 values thus making it difficult to discriminate between the various pathological states[2].

Fundamental research into the structure and function of cancer cells by NMR spectroscopy has produced both fruitful and disappointing results. Despite enormous effort its contribution has been significantly smaller than anticipated.

Why has NMR not been as productive in cancer research as in other disciplines? Undoubtedly the single most important feature was a lack of understanding of the biology of the cancer cell. The realisation that cancer cells have the capacity to diversify and create cellular variants has become apparent only in the last few years[3]. Even if a cancer originates from a single transformed cell, a tumor cell population at the time of clinical manifestation will be phenotypically and genotypically heterogeneous. Maintenance *in vivo* or *in vitro* of experimental models is very difficult and undertaking NMR experiments on samples several weeks apart could yield different results.

Secondly, it has been assumed that the physical and chemical properties of normal cells would be sufficiently similar to the cancer cell to allow the same experimental methods to be used. We will demonstrate that this is not always the case and suggest that the biophysicist undertaking NMR studies on cancer cells should make no assumptions based on the nature of healthy cells. Furthermore the NMR data are useful only if the biological status of the cells is verified by biochemical, chemical, animal and flow cytometric analysis.

The magnetic resonance experiment cannot be explained adequately without recourse to classical physics or quantum mechanics[4]. We shall not attempt to explain these experiments; we shall outline them briefly and direct the reader towards the NMR literature. However, we shall describe in detail the type of information that this very powerful technique can provide on this proliferative and usually terminal disease, cancer.

B. NMR Techniques Suitable for the Study of Cancer Cells

I. High Resolution NMR Spectroscopy

The NMR spectra obtained from small molecules in non-viscous solvents comprise narrow lines, often one group for each chemically distinct site in the molecule. Figure 1 shows a high resolution ^{1}H NMR spectrum of triolein in chloroform, with assignments. The characteristic frequency of a particular line in the spectrum is usually expressed as the *chemical shift* in parts per million (ppm). The multiplicity of some of the lines centred around the chemical shift values is due to interaction between the spins themselves. These so-called *spin-spin coupling constants* occur over only a few bonds, and thus are helpful in characterising the atoms neighbouring that of interest e.g. the protons of the terminal methyl group are coupled to the two protons of the adjacent methylene giving rise to a triplet (Fig. 1)[5].

In addition to ^{1}H, there are many other magnetically active nuclei. A partial listing, with relevant properties, is given in Table I. Note that as the NMR frequency decreases, so does the detection sensitivity. Low isotopic abundances can make nuclei such as ^{13}C and ^{15}N rather difficult to detect. On the other hand, isotopic enrichment in these elements offers a means to observe particular compounds with little interference from a natural abundance background.

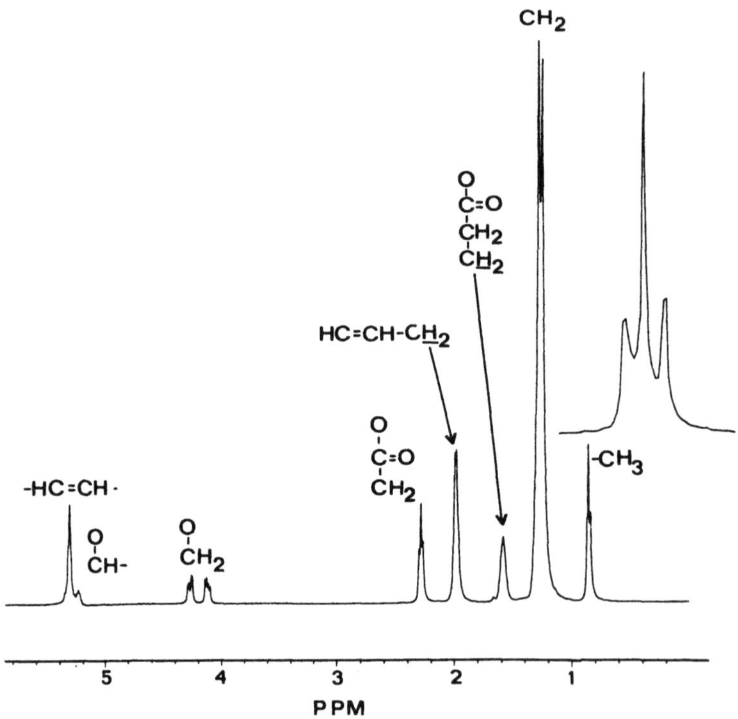

Fig. 1. 400 MHz ^{1}H NMR spectrum of triolein in $CDCl_3$. The inset is an expansion of the region from 0.77 to 0.92 ppm

The NMR experiment involves excitation of a nuclear spin system. After excitation the system returns to equilibrium – a process known as relaxation. Depending upon the type of experiment, several types of relaxation can be measured. We shall confine ourselves to the two more common types.

In the presence of a magnetic field, and the absence of electromagnetic excitation, more nuclear spins will be orientated with their magnetic moments parallel to the applied field than antiparallel, resulting in a net magnetisation of the ensemble of spins. Excitation with radio frequency radiation decreases the component of this magnetisation *parallel* to the field direction; the return towards the unperturbed value is known as *longitudinal,* spin-lattice, or *T_1 relaxation.* The quantity T_1 is in fact the time constant characterising this relaxation.

Excitation of the spin system causes a large number of spins to precess in synchrony. This leads to a significant component of magnetisation *perpendicular* to the applied magnetic field. On removal of the exciting pulse, desynchronisation of the precessions of individual spins begins, leading to diminution of magnetisation in the perpendicular direction. This process is known as *transverse,* spin-spin or *T_2 relaxation.* A significant contributor to this process may be the inhomogeneity of the static magnetic field; measurements made without compensation for this lead to an effective relaxation time T_2^*. However, there are methods available to remove this contribution, and thus to measure the intrinsic T_2 values.

As there are a variety of mechanisms contributing to both T_1 and T_2 relaxation, the terms spin-lattice and spin-spin relaxation, which were coined in the early days of ^1H NMR, are less preferred than the operational descriptions, *longitudinal and transverse relaxation.*

The relaxation times T_1 and T_2 are measured by particular sequences of radio frequency pulses[6]. For T_1 determination, a simple sequence is (π)-τ-$(\pi/2)$-Acquire. The (π) represents a pulse of 180° which inverts the magnetisation to become effectively antiparallel to the static field. During a waiting period of τ seconds, this magnetisation relaxes back via T_1 processes towards the equilibrium value. The $(\pi/2)$ or (90°) pulse rotates remaining longitudinal magnetisation into the transverse direction where a signal may be detected during the acquire time. The dependence of this signal amplitude on the time τ leads to a simple measure of T_1.

Transverse relaxation times T_2 are more subject to errors in measurement. A reliable method, known as the CPMG (Carr-Purcell-Meiboom-Gill) applies a sequence $(\pi/2)_x$-τ-$[(\pi)_y$-τ-Echo$]_n$. The $\pi/2$ and π pulses are applied with phase shifts of 90° with respect to one another, hence the subscripts x and y. At a time τ after the π pulse an echo signal appears in the transverse direction. If n π pulses are applied, n corresponding echo signals will be generated, whose decay with increasing n-value is a measure of T_2. By using only the even echoes, compensation for errors in pulse length is achieved. The sequence is still sensitive to drifts in frequency and care must be paid to spectrometer adjustment in general.

The *spin echoes* described above can be useful in a completely different manner. In living cells or tissues there is a wide variety of compounds of differing sizes and relaxation times. Usually the larger the molecule, and the less its mobility, the shorter will be its T_2 values. Recall that the NMR signal is derived from the magnetisation in the plane transverse to the applied magnetic field, which decays according to T_2. Thus, magnetisation due to large molecules such as DNA, collagen, etc., will

Table I. Nuclei of potential interest for NMR studies of cancer cells

Nucleus	NMR frequency at 2.35 T (MHz)	Natural abundance (%)	Sensitivity relative to ^{1}H
^{1}H	100	99.98	1.00
^{19}F	94.1	100.0	0.83
^{31}P	40.5	100.00	6.63×10^{-2}
^{23}Na	26.5	100.00	9.25×10^{-2}
^{13}C	25.1	1.11	1.59×10^{-2}
^{2}H	15.4	0.015	9.65×10^{-3}
^{15}N	10.1	0.37	1.04×10^{-3}
^{39}K	4.67	93.1	5.08×10^{-4}
^{41}K	2.56	6.88	8.40×10^{-5}

decay rapidly relative to that due to cellular metabolites such as amino acids or triglycerides. By adjusting the internal τ in the pulse sequence $(\pi/2)$-τ-(π)-τ-Echo, it is possible to minimise the contribution from large molecules and observe signals due only to compounds of long T_2[7].

In proton NMR, suppression of the resonance due to H_2O is a major problem. Without some means of suppression, this resonance can dominate the spectrum by factors of thousands (hydrogen in water is 110 M!). A variety of methods exists.

The simplest method to reduce the intensity of the water peak is to replace H_2O by heavy water (D_2O). Although deuterium has a magnetic moment, it resonates at a very different frequency from that of ^{1}H (Table I).

When the heavy water technique is undesirable, or ineffective, various irradiation methods may be used to minimise the height of the H_2O resonance. The simplest is to irradiate continuously, for a time long relative to the T_1 of H_2O, at the frequency of the H_2O resonance, before beginning the regular pulse sequence. This equalises the populations of spins whose moments are parallel and antiparallel to the applied field, causing the H_2O signal to tend towards zero intensity. A problem with this technique is that it may cause aberration of resonances close to H_2O in the ^{1}H NMR spectrum. Thus, another approach is to irradiate the H_2O with "soft" pulses, i.e. pulses of long duration and weak intensity, whose excitation region is very narrow[8]. There are more complicated ways to achieve this, such as the pulse sequences due to Redfield[9]. Finally, one may take advantage of the fact that the T_1 of water is often very different from those of the compounds of interest[10]. The T_1 measuring sequence (π)-τ-$(\pi/2)$-Acquire is used, with τ carefully adjusted such that the amplitude of the H_2O resonance is exactly zero (due to the use of a π pulse, short values of τ lead to spectra of negative amplitude, whereas long values of τ lead to positive amplitudes).

II. Broad Line NMR Spectroscopy

Until recently published NMR spectra contained mainly narrow lines resonances. This is because most of the interest was in small molecules dissolved in solvents of low viscosity. It was known that large molecules give broad resonances, and that solids yield very broad spectra, but the early high resolution spectrometers had difficulty in observing either kind. In the present generation of instruments there are spectrometers that observe any or all types of spectra.

Why are the resonances of large molecules or solids so broad? The problem lies in the fact that most NMR parameters depend upon the angle between a particular direction in the molecule and the applied static magnetic field. For example, the chemical shift of ^{31}P due to the phosphodiester moiety in a phospholipid can vary from + 80 to − 110 ppm depending on this angle – the chemical shift is highly anisotropic[11]. However, in non-viscous solution rapid molecular motion averages the chemical shift, the resulting value being the algebraic average of the shifts in the three principal directions. Similar arguments apply to the widths of the resonances, which in large molecules or solids may be fairly broad due to the sum of a large number of anisotropic interactions of various magnitudes. In large molecules, or large assemblies such as membranes or ribosomes, the overall rate of motion is often

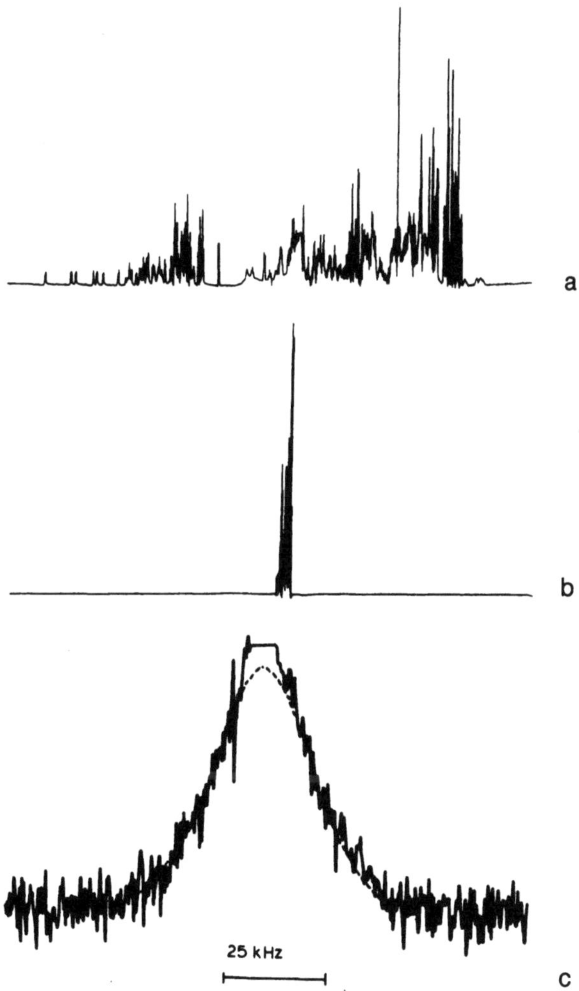

Fig. 2. Comparison of the ^1H NMR spectra of proteins in dilute aqueous solution and in slowly tumbling membranes. **(a)** The 400 MHz ^1H NMR spectrum of bovine pancreatic trypsin inhibitor, in aqueous solution, exhibits a very rich high resolution spectrum with an enormous number of sharp lines visible over a range of a few KHz. **(b)** Spectrum of **(a)** on the scale of **(c)**. **(c)** The ^1H NMR spectrum of rhodopsin reconstituted in model membranes of dimyristoylphosphatidylcholine with perdeuterated acyl chains exhibits a broad featureless line approximately 25 KHz wide. In **(c)** all the fine structure is masked by incompletely averaged proton-proton dipolar interactions[12]

too slow to average these interactions effectively, and the resonances can be rather broad. For example, even though there is some residual motion of phospholipids within the bilayer structures of membranes, their ^{31}P NMR spectra are usually of width 35–50 ppm[11]. A further, and more dramatic example of this behaviour, is seen in the ^{1}H NMR spectra of Fig. 2, where the magnificent detail obtainable from the high resolution spectrum of a small peptide is contrasted with the shapeless lump obtained from a protein in a membrane-protein complex[12].

Despite the breadth of the resonances and spectra obtained from large structures such as membranes, an impressive amount of detail is actually obtainable on the organisation and mobility of the component lipids. The types of experiment and analysis are described in reviews on the use of ^{31}P[11, 13], ^{2}H[12–14], and ^{1}H[15, 16] in membrane research.

III. Magnetic Resonance Imaging

During the past five years the use of NMR techniques for anatomical imaging has grown exponentially. The methods used are very different from those described here so far, and the data obtained yield essentially structural rather than chemical images, although effective T_1 and T_2 values can be measured. An impressive number of books on the subject is already available[17–19].

Most NMR imaging, now known as MRI to avoid any mistaken associations with radioactivity, either malignant or therapeutic, involves ^{1}H NMR, and essentially measures proton density. Radio frequency, special pulse sequences, and rapidly changing magnetic field gradients are employed. The result can be a slice through a given piece of anatomy, with resolution as good as 0.1 mm. With higher magnetic fields much higher detection sensitivity is gained, but this is concomitant with an increasingly troublesome image artifact due to the different chemical shifts of the various chemical species within the anatomy. However, since the principal contributors to the proton signal are fat and water, it has become possible to produce images due uniquely to fat or water, yielding even better contrast.

In parallel with the surge of activity in MRI has come the possibility of performing spectroscopic measurements (chemical shifts, specific T_1 and T_2 values, biochemical kinetics) on intact bodies via ^{1}H, ^{31}P and ^{13}C NMR. Much success has already been achieved with animals[20–22] and peripheral limbs of humans[23, 24]. Spectra of ^{31}P in human brain have been reported[25, 29]. However, in all the above the problem remains of knowing exactly where one is looking in the anatomy. Recently some techniques combining the methods of imaging and spectroscopy have been reported which hold promise for resolving the problem of spatial specificity[26, 27].

C. High Resolution NMR of Intracellular Compounds in Cancer Cells

I. ^1H NMR

The small molecule metabolites within cells should yield narrow NMR resonances unless they are tightly bound to large molecules or superstructures. They should therefore be visible to conventional ^1H NMR spectrometers. This they are, but the large number of them and the relatively small range of ^1H NMR chemical shifts makes it difficult to distinguish one species from another. With the greater availability of high field instruments and their improved detection sensitivity and spectral dispersion, it is now possible to separate and assign individual resonances by ^1H NMR[7)].

Agris and Campbell have used ^1H NMR at 470 MHz to follow the metabolites in Friend leukaemia cells during their erythroid-like differentiation[28)]. Addition of dimethylsulfoxide led to the production of globin and haemoglobin. Using the spin echo technique described earlier they were able to follow the time dependence of a variety of resonances. Resonance assignment was made by comparison with the corresponding spectra of cell extracts; in all, some 64 resonances were assigned to 12 amino acids and 19 compounds involved in intermediary metabolism. The most dramatic effect seen during the differentiation was a four-fold increase in the concentrations of glycerophosphorylcholine and phosphorylcholine. A concomitant increase in resonances attributed to the methylene groups of triglyceride was also reported.

^1H NMR of the water in cancer cells was one of the first applications of the NMR technique in oncology[1)]. Early reports suggested that cancerous tissue had T_1 and T_2 values for water significantly longer than those of normal tissue[1,43)]. Subsequent studies pointed out that cancerous tissue contained increased amounts of water which could be responsible for the longer T_1 values[44)]. As time has gone on, it has become clear that the correlations between water relaxation parameters and pathological states are at best vague, and that significant differences in sample preparation and methods of measurement make comparison of results difficult. NMR of water in biological systems was reviewed recently by Bottomley and co-workers[41)] and a review on the use of NMR of water in pathological tissue is covered in a second review[2)]. An historical overview was presented recently[73)].

II. ^{31}P NMR

Phosphorus-31 NMR was used to visualize the metabolites in a tumour within a rat, using surface coil detection and magnetic field focussing to observe the tumour uniquely[21)]. The flat surface coil[20)] was placed on the surface of the animal. This coil configuration detects mainly the tissue below the coil. To ensure that only the tumour, and not also the muscle between the tumour and the surface coil was observed, the magnetic field was adjusted so as to be unfavourably inhomogeneous everywhere but in the region of the tumour. This procedure is known as topical NMR.

In this initial *in vivo* ^{31}P experiment on a tumour, a Walker 256 carcinosarcoma was studied. The pH of the tumour was compared with that of the neighbouring muscle by means of the chemical shift of inorganic phosphate, and found to be slightly lower (7.14 compared to 7.21). Infusion of glucose into the animal did not yield the anticipated decrease in pH. Deoxyglucose was infused, and visualized in the form of deoxyglucose-6-phosphate. Estimation of pH from this resonance yielded a different value (6.44) from that measured from inorganic phosphate (7.14). On the other hand, in the ischaemic tumour the pH values measured by both probes were identical (6.32 and 6.33). The different pH values measured in the unaltered tumour by the two probes were attributed to compartmentalisation of the probes in different regions of the tumour.

Subsequently Griffiths et al. demonstrated the viability of this technique for the study of ^{31}P in human subjects, using a tumour (Alveolar rhabdomyosarcoma) on the hand[30)]. The tumour showed a lower than usual ratio of intensities of resonances due to phosphocreatine (PCr) and adenosine triphosphate (ATP). Progression in the tumour after unsuccessful chemotherapy led to a further decrease in this ratio. These observations were in accord with those of Ng et al. who found that successful chemotherapy led to increases in this ratio[31)].

In a subsequent study of mammary 16/C tumours implanted subcutaneously in mice, Evanochko et al. followed in detail the response of metabolites during untreated growth and after treatment with adriamycin, hyperthermia, or X-radiation. With increase in tumour mass, the levels of PCr and ATP decreased, that of inorganic phosphate (P_i) increased, and the apparent pH as measured from the P_i resonance decreased, Fig. 3[32)]. These variations were attributed to increasing hypoxia due to changes in tumour vascularisation. Partial reversal of these changes was observed during response of the tumour to chemotherapy. Hyperthermia led to a ^{31}P spectrum characteristic of dead tissue with high levels of P_i, low levels of ATP, and acidic pH. After X-radiation the tumour receded, and the ^{31}P NMR spectrum changed in the direction of that of a more aerobic tumour, with increasing levels of phosphocreatine.

A very detailed study of the metabolites of radiation-induced fibrosarcoma tumours grown in mice was performed by Evanochko et al. using ^{1}H, ^{13}C and ^{31}P NMR and various solution chromatographic methods[33)]. Tumours were extracted after freeze-clamping and the high resolution spectra of the extracts were compared with those of the intact tumour. The procedures allowed accurate assignments of the various NMR peaks. The resonances in the in vivo ^{31}P spectrum usually attributed to sugar phosphates were found to be due to other phosphomonoesters (PME) such as phosphorylcholine. Phosphocreatine was found to occur definitely in the tumours as well as in surrounding tissue. In the ^{13}C NMR spectra of tumour extracts, peaks due to lactic acid and the aminosulfonic acid taurine were found.

The influence of hyperthermia on the levels of high energy phosphates of Dunn osteosarcoma tumours in mice was investigated by Lilly and co-workers[34)]. The degree of fall of the ATP/P_i ratio was directly proportional to the heat dose and to the thermal cell kill. It was suggested that the ^{31}P method could be useful in following the efficacy of hyperthermia as an antitumour modality.

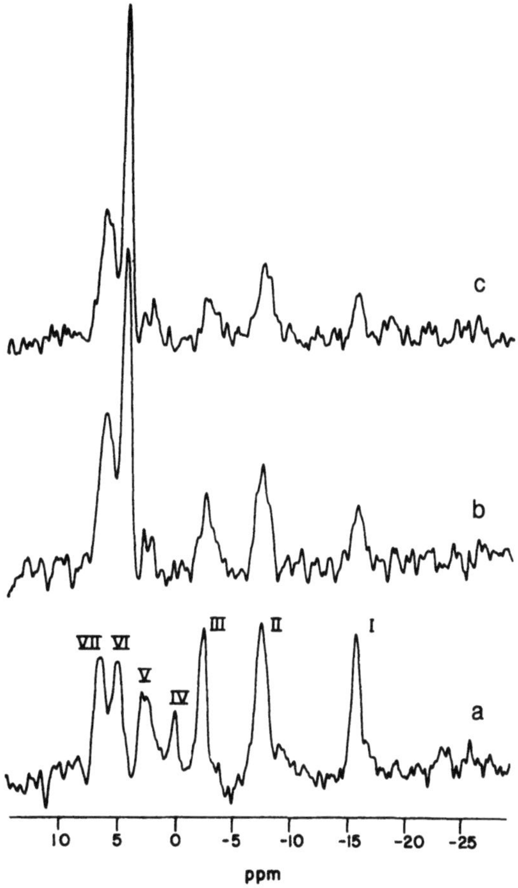

Fig. 3. ^{31}P NMR (81 MHz) spectra of a mammary adenocarcinoma, implanted subcutaneously in a mouse, taken at different times after implantation. **(a)** 9 days, tumour mass 2 g, pH = 7.4. **(b)** 10 days, mass 3.5 g, pH = 7.0. **(c)** 11 days, mass 5.0 g, pH = 6.8. The peaks are labelled in spectrum **(a)**: I, ATP_β; II, ATP_α and ADP_α and NAD; III, ATP_α and ADP_β; IV, PCr; V, unidentified, VI, inorganic phosphate; VII, sugar phosphate[32)]

The first NMR study of an endocrine tumour was reported by Prysor-Jones et al.[35)]. They used a surface coil to follow the phosphorus-containing metabolites in pituitary tumours implanted in rats. They showed increases in levels of inorganic phosphate, and a decrease in phosphocreatine and the apparent pH, on treatment of the rat with the prolactin secretion-stimulating peptide, vasoactive intestinal polypeptide. Thyrotrophin-releasing hormone caused increased in P_i, but no change in pH. These changes were concomitant with the release of prolactin from the endocrine tumour.

Most recently, the ^{31}P NMR surface coil technique has been applied to the study of neuroblastoma in two human neonates by Maris et al.[36)]. One child suffered from a mild, naturally regressing form of the disease (Stage IV-S) whereas the other was afflicted with a form involving skeletal and distant lymph node metastases (Stage IV) for which aggressive therapy was essential. Both babies had enlarged livers at first presentation, and in the ^{31}P spectra of both the PME peaks were of increased intensity relative to those in the spectra from healthy liver (see PME peaks in Fig. 4). Six weeks later, the liver of the baby with naturally regressing disease had levels of PME almost equal to those of normal livers, whereas the Stage IV baby's liver yielded PME levels elevated by seven times (Figs. 4B and C). However, four

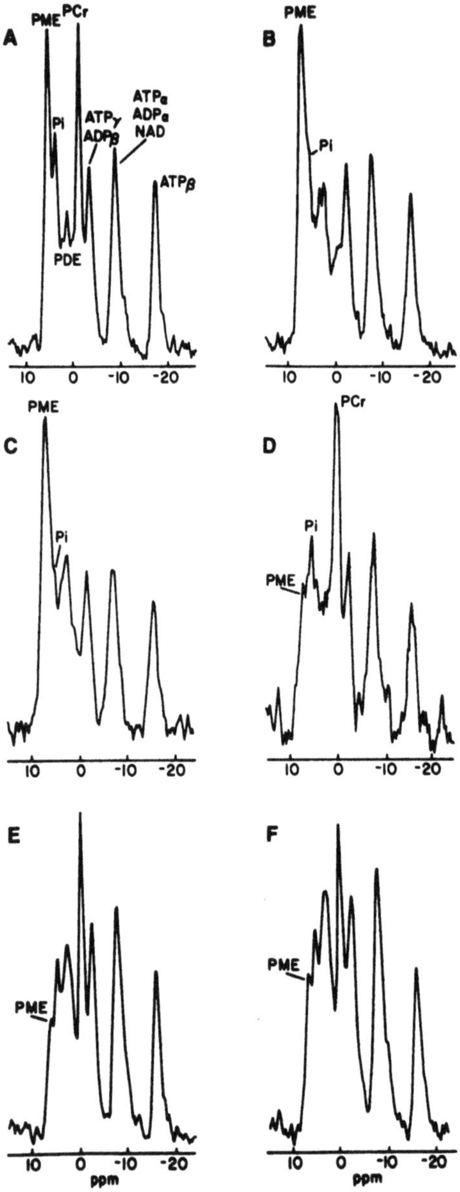

Fig. 4. ^{31}P NMR (33 MHz) spectra of the liver of a female patient suffering from Stage IV neuroblastoma, taken as a function of time. **(a)** At first presentation, patient age *ca.* 4 weeks. **(b)** Four weeks later, 10 days after X-irradiation. **(c)** Eight weeks later. **(d)** 16 weeks after initial response to treatment which was begun at age 7 weeks. **(e)** 24 weeks after initial response to treatment, coil at edge of liver. **(f)** As in **(e)**, coil anterolateral to edge of liver. Abbreviations, other than those already cited, are: PME = phosphomonoester; PDE = phosphodiester[36)]

and six months after response to treatment and improvement of symptoms (Figs. 4D and F) the PME signal of the stage IV patient had decreased towards the value of the control livers. High resolution NMR studies of biopsy material showed that the PME peak was due to phosphorylethanolamine at a concentration of 10 mM.

This study yielded four significant conclusions. Firstly, tumour tissue can be distinguished from host tissue *in situ,* since neuroblastoma has a characteristic ^{31}P spectrum with high PME : ATP_{β} ratios. Secondly, the PME spectra of the rapidly enlarging and the spontaneously regressing livers were quantitatively different in their PME : ATP_β ratios. Thirdly, the time-dependent changes in the ^{31}P NMR spectra correlated well with the clinical observations. Finally, the incidence of high levels of phosphorylethanolamine in neuroblastoma correlates well with the high levels of PME found in human neonatal brain and developing dog brain. The neonatal brain is the only other human tissue in which PME : ATP_β ratios greater than one have been found to date, perhaps indicating a need for high rates of phospholipid synthesis.

The responses to chemotherapy of tumours in perfused human kidneys have been followed by ^{31}P NMR[92]. The tumour region of some of the kidneys yielded a resonance at 4.2 ppm which was not seen in normal kidneys or when the surface coil was over non-tumour-bearing regions. It was concluded that the 4.2 ppm resonance was due to inorganic phosphate in a compartment of low pH (6.10–6.55). In other respects the ^{31}P NMR spectra of the tumours were relatively normal and indicative of a high rate of metabolic activity. Furthermore, the response of the tumours to hypoxia was much less than that of normal kidneys. The peak at 4.2 ppm was absent from the ^{31}P spectra of the tumours under hypoxia or ischaemic perfusion. A kidney treated with doxorubicin (100 mg/ml) for 4 h showed no change in the spectrum of either tumour or normal tissue. However, another kidney, infused with a mixture of doxorubicin, actinomycin-D, cyclophosphamide, and vincristine, showed a marked increase in intensity of the peak at 4.2 ppm and a decrease in the level of ATP, in the region of the tumour. The normal region of the same kidney showed no such changes during chemotherapy.

Recently a report appeared on the observation of human prostate tumours, induced in mice, by the use of ^{31}P NMR[93]. The spectra were responsive to treatment with X-rays, both in PCr levels and in intracellular pH. Several other reports of the responsiveness of the ^{31}P spectra of tumours to chemotherapy or radiation have appeared[83, 86, 89].

III. ^{19}F NMR

Although ^{19}F has NMR characteristics very favourable for detection and chemical identification, it is rarely used in biological studies because no endogenous compounds contain fluorine. To make use of it compounds must be labelled specifically. An obvious candidate for ^{19}F NMR studies of cancer is the fluorine in various anticancer drugs. Recently two such studies have been reported[37, 38].

The wide range of ^{19}F chemical shifts was used by Malet-Martino and co-workers[38] to follow the catabolism of 5′-deoxy-5-fluorouridine in the blood, serum and urine of a patient undergoing chemotherapy. The absence of natural fluorine-containing samples dictated that any ^{19}F signal must come from the drug. Resonances assignable to a variety of compounds were detected, including an unexpected metabolite, α-fluoro-β-alanine. Some ^{19}F spectra from the study are shown in Fig. 5. Note the ^{1}H-^{19}F spin-spin coupling multiplets in the spectra. The large couplings of *ca.* 50 Hz are due to, and characteristic of, ^{1}H-^{19}F in the –CHF– fragment of the 5,6-dihydrofluorouracil moiety. The major metabolites found in blood were 5-fluorouracil, 5,6-dihydrofluorouracil and α-fluoro-β-alanine. In urine the principal components containing fluorine were the unmetabolized drug and the alanine derivative.

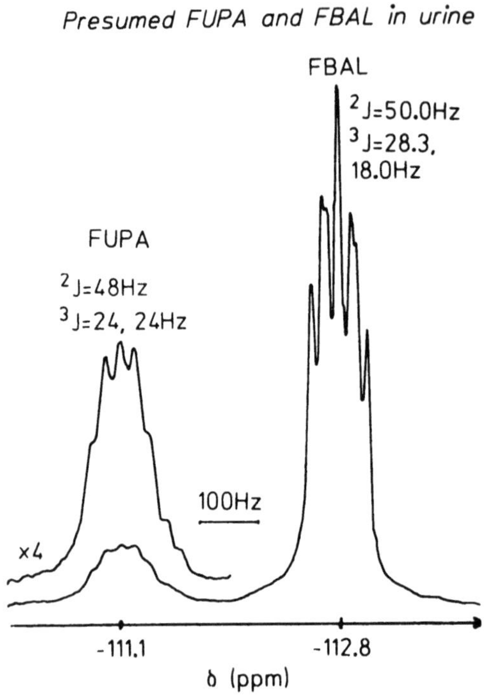

Fig. 5. ^{19}F NMR (250 MHz) spectra of α-fluoro-β-alanine (FBAL) and presumed α-fluoro-β-ureido-propionic acid (FUPA) in urine[38)]

The time dependence of the concentrations of these compounds was determined. Recently these studies have been carried out in greater detail on urine, to yield a complete profile of metabolites of 5-fluorouracil[39)].

Stevens et al. applied the surface coil method to study the metabolism of 5-fluorouracil (5 FU) in animal-borne tumours and in the livers of mice afflicted with Lewis lung carcinoma[37)]. The data demonstrated that in liver 5 FU is degraded to dihydrofluorouracil and α-fluoro-β-alanine. No toxic nucleosides or nucleotides were observed in the liver. The tumour tissue degraded 5 FU more slowly than the liver, and the catabolic products found in liver were not detected. Deoxynucleoside and nucleotide derivatives were evident in the tumours.

The ^{19}F method appears to have great utility in elucidating the pharmacology and pharmacokinetics of fluorinated anticancer drugs. The high detection sensitivity bodes well for its use on human subjects.

D. The Plasma Membrane of Cancer Cells Gives a High Resolution Lipid Spectrum

I. ^1H NMR Spectra of Cancer Cells

The ^1H NMR spectrum of an excised distinct solid tumour is remarkably similar to that obtained from a suspension of the same type of cells grown in culture[45] (Fig. 6). This spectrum is characteristic of lipid molecules, and can be enhanced in resolution by Lorentzian-Gaussian deconvolution methods to expose at least four resonances under the broad methylene peak at 1.2 ppm[45] (Fig. 6C). It is these four resonances which provide information on the cells' ability to metastasise or reject drug therapy[46–48]. Similar spectra have been reported recently for lymphocytes of human patients with chronic lymphatic leukaemia[66].

Block and co-workers, who first reported the existence of a high resolution NMR spectrum, compared the intact EL4 cell line with extracted cellular lipid. Based on

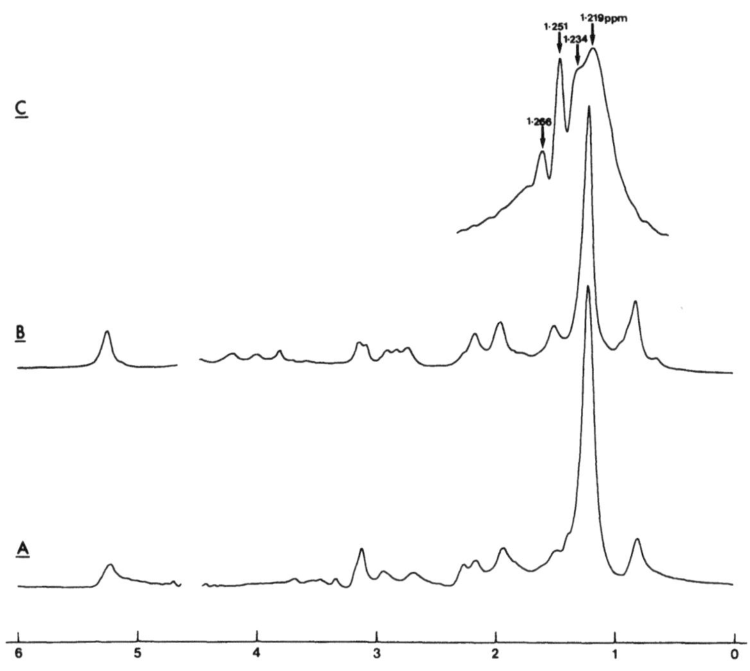

Fig. 6. 400 MHz ^1H NMR spectra of the mammary adenocarcinoma cell line J clone in Ca^{2+} and Mg^{2+} containing phosphate-buffered saline in D_2O. Data were recorded with the sample spinning, 37°. The water peak was suppressed by gated irradiation[91]: 4000 Hz sweep width, 1.645 s acquisition time, 64 accumulations. **(A)** A solid tumour excised from a Fisher rat. The 90° pulse width was 9.5 μs and a line broadening of 3 Hz was applied. **(B)** "J clone" cells (1×10^8) were suspended in 0.4 ml of above buffer, pulse width was 8.5 μs, and a line broadening of 3 Hz was applied. **(C)** A Lorentzian-Gaussian resolution enhancement ($k = -11$ Hz, $g = 0.04$ and $a = 1.64$ s) was applied to the free induction decay recorded for the J clone sample above. The four resonances resolved under the $-CH_2-$ are plotted on an expanded scale

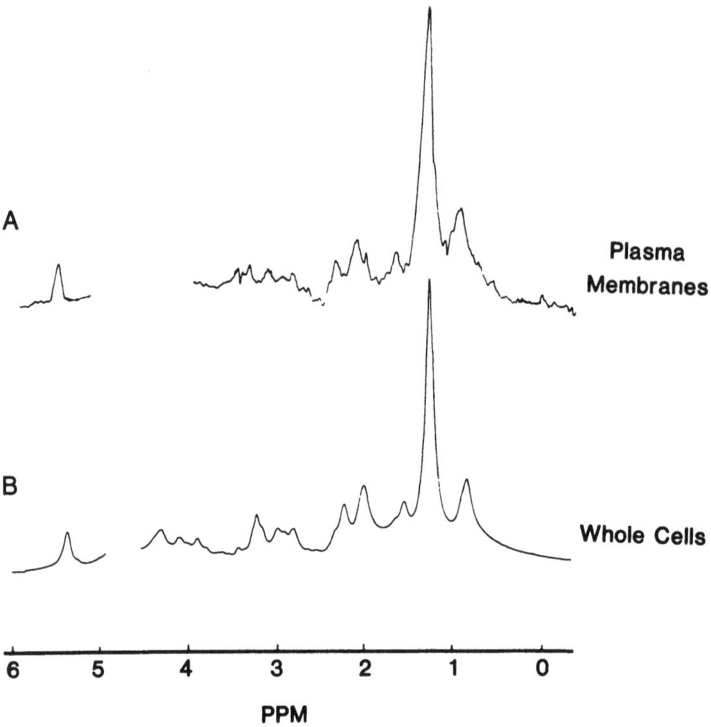

Fig. 7. ^{1}H NMR spectra (400 MHz). **(A)** Membrane ghosts (2 × 10^7) isolated from the EBV cell line JP resuspended in PBS in D_2O. **(B)** EBV cell line JP (5 × 10^7) suspended in PBS in D_2O. Spectra were recorded in 11 min at 37° on a Bruker WM-400 NMR spectrometer with the residual HOD peak suppressed by selective irradiation[91)]. A line broadening of 2 Hz was applied

these data they suggested that non water ^{1}H NMR signals from tissue may be able to reflect some pathologically relevant information[49–52)]. Other reports of a high resolution lipid spectrum include non transformed cells such as fibroblasts[53–55)], peripheral blood B lymphocytes, and stimulated T and B lymphocytes[56, 57)].

Narrow lined ^{1}H NMR lipid spectra from intact cells and tumours were considered by most as unlikely, since to generate such a spectrum the lipids had to be able to tumble independently of the cell or tumour. Most scientists familiar with the NMR technique considered that such signals could only be from liposomes or fat droplets in the cytoplasm and therefore would be of little relevance to cellular mechanisms.

Several years later, and with the advantage of spectrometers operating at higher field strengths, we were able to confirm that these signals are indicative of the cells' biological capacity[46–48)]. In addition, we have attempted to identify the source of the NMR signal.

II. The Location of the Informative NMR Signal is the Plasma Membrane

The protons responsible for the narrow NMR lines were found to be lipids either in or associated with the plasma membrane.

1) Isolation of Plasma Membrane Ghosts. The first direct evidence that plasma membrane lipid was responsible for the cellular ^1H NMR spectrum was obtained by comparing isolated membrane ghosts with intact cells. Spectra of the plasma membrane ghosts[57)], and of lipid isolated from a plasma membrane preparation enriched four-fold[46)], were directly comparable with that of the intact viable cells (Fig. 7).

2) Paramagnetic Probes. The paramagnetic metal ions Gd^{3+} and Mn^{2+} broaden resonances in the NMR spectra of nearby nuclei[58)]. Gadolinium chloride, when titrated into a suspension of the leukaemic T-cell line CCRF-CEM (in $NaCl/D_2O$), selectively broadens resonances in the spectrum, including the methylene resonance at 1.2 ppm (Fig. 8A). Manganese ions, on the other hand, broaden resonances in the cell spectrum non-selectively (Fig. 8B).

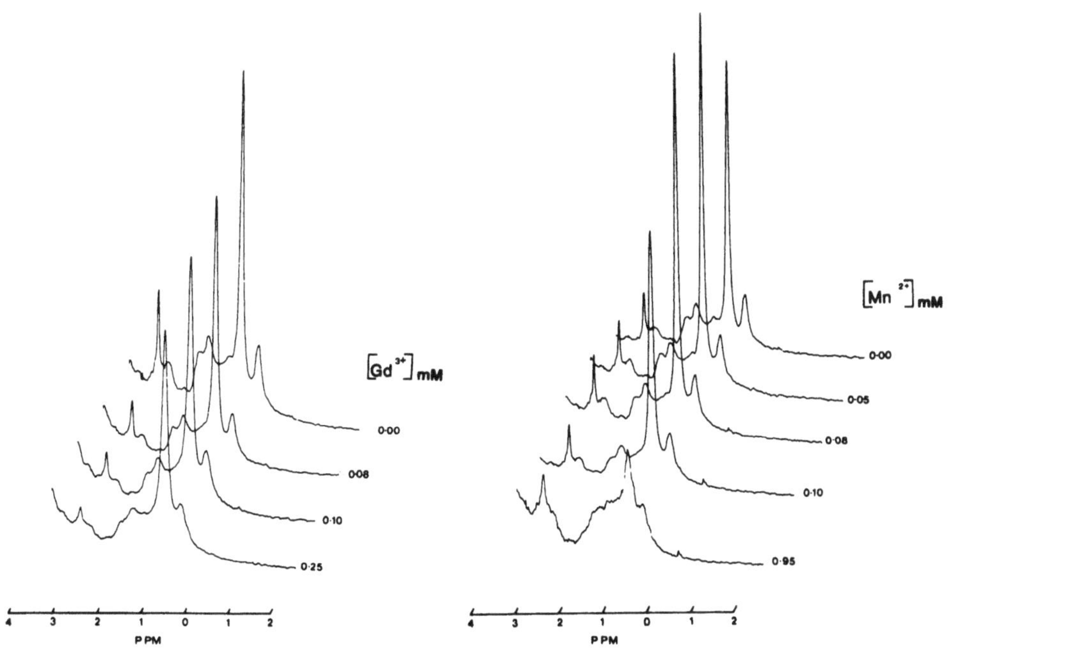

Fig. 8. ^1H NMR spectra (100 MHz) of leukaemic T-cell line CCRF-CEM (5×10^7 cells) in $NaCl/D_2O$. **(A)** With increasing amounts of $GdCl_3$. **(B)** With increasing amounts of $MnCl_2$. Spectra were recorded at 37° in 17 min on a Jeol FX-100 NMR spectrometer. The residual HOD peak was suppressed by selective gated irradiation[91)]

Table II. Location of isotopes after incubation with leukaemic T-cell ghosts

	Location of isotopes (%)	
	Plasma membrane	Inside the cell
^{153}Gd	92	8
^{154}Mn	34	66

Ghosts were prepared from the leukaemic T-cell line CCRF-CEM[56)], and the sample was divided into two. Either $^{54}MnCl_3$ or $^{153}GdCl_3$ was incubated with the ghosts for 10 min at 37°, and the sample was washed twice. The ghosts were lysed, and the supernatant was drawn off.

To determine if the cellular lipid affected by these paramagnetic probes was part of the plasma membrane intact membrane ghosts were exposed to either $^{153}Gd^{3+}$ or $^{54}Mn^{2+}$. The location of these radioactively labelled metals indicated that $^{54}Mn^{2+}$ was able to pass through the plasma membrane whereas the $^{153}Gd^{3+}$ selectively bound to the membrane (Table II).

These experiments parallel those described by Bergelson and Barsukov[58)], where Gd^{3+}, unable to pass through the vesicle bilayer broadened only the external lipid molecules. In contrast Mn^{2+}, able to pass through the bilayer, broadened all resonances.

More detailed structural information is obtained if the spectra are collected at 400 MHz and resolution enhancement techniques invoked in conjunction with the Carr-Purcell Meiboom Gill pulse sequence[45, 47)]. Of the four resonances under the methylene envelope at 1.2 ppm, Gd^{3+} broadens the one at 1.22 ppm, partially broadens that at 1.25 ppm, whereas the third resonance, at 1.23 ppm, is only slightly affected by the presence of the paramagnetic metal ion (Fig. 9). The fourth resonance at 1.27 ppm is also affected.

III. A Physical Explanation for the Narrow Resonances

Most lipid molecules in the plasma membranes of cells should generate a ^1H NMR spectrum of width about 10 KHz, even when the membranes are in the liquid-crystalline state[15, 60, 62)], due to the lack of averaging of dipolar interactions to zero. In contrast, the narrow lines originating from the plasma membranes of cancer cells are less than 10 Hz wide.

There are two potential mechanisms that can account for narrow lines associated with the protons in this system. The first involves lipid molecules located at positions in the plasma membrane such that the principal axis of motional averaging makes an angle θ with the external magnetic field close to the "magic angle". At this angle the rapid reorientational motions and conformational transitions are capable of averaging the intra-molecular dipolar interactions to zero.

The second possibility is that the narrow lines are quasi-Lorentzian in nature[59, 61)], and arise from small structures which retain their integrity within the plasma membrane, or closely associated with it, but allow molecules within them to

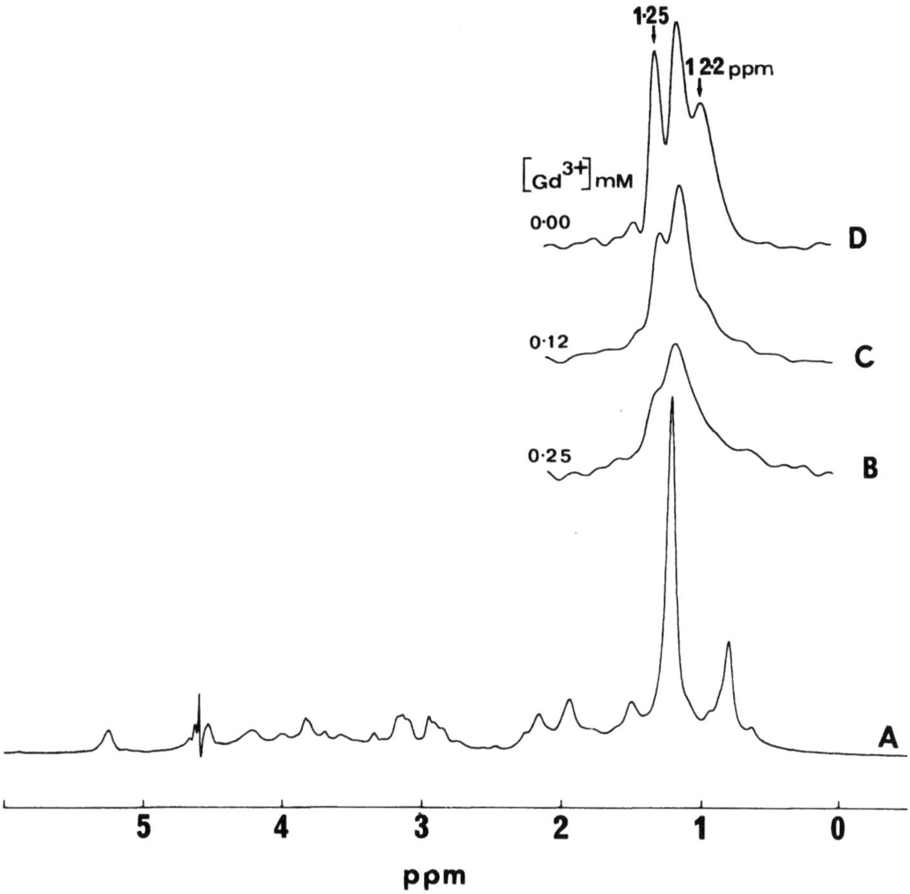

Fig. 9. ^{1}H NMR (400 MHz) spectra of (**A**) 13762 cells (1×10^8 cells/400 μl) suspended in PBS/D_2O; (**B**), (**C**) and (**D**), are expansions of the methylene resonance after Lorentzian-Gaussian resolution enhancement[45]. (**B**) and (**C**) have 0.12 and 0.25 mM Gd^{3+} added, respectively

tumble isotropically and produce narrow resonances. A requirement imposed by the theory of motional narrowing[4] for the existence of such distinct resonances is that the molecules of these small structures should not exchange appreciably with the membrane lipid molecules in a time $\leq T_2$ = hundreds of milliseconds.

To test these two possibilities selective and non-selective T_1 experiments were undertaken on a suspension of cells[45]. The non-selective inversion recovery experiment showed single exponential behaviour for up to 2 s with a T_1 of 600 ms. A further experiment which selectively involved the methylene resonance at 1.2 ppm gave a T_1 of 510 ms. Similar results were obtained with solid tumours yielding non-selective and selective T_1 values of 690 and 550 ms respectively[45].

A rigorous lower limit to the time required for diffusive exchange between the molecules contributing to the narrow ^{1}H NMR line and those of the broad underlying spectrum is provided by the observation that selective and non-selective inversion-recovery T_1 measurements give very similar results. This indicates that no

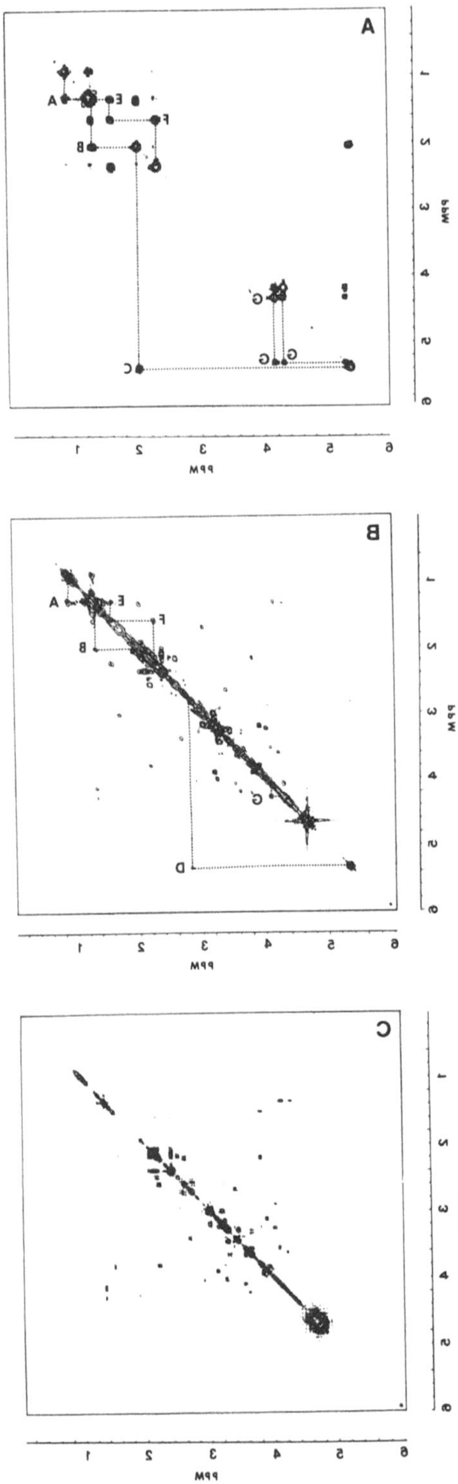

Fig. 10. 400 MHz symmetrised COSY spectra collected at 37°, as previously described[65]. **(A)** Triolein in $CDCl_3$ (20 mg/ml). **(B)** Human leukaemic T lymphocyte cell line VBL 20 (1×10^8/400 µl PBS/D_2O). **(C)** Supernatant containing only cytoplasmic components after cell membranes, nuclei, etc. are spun down. The assignments of the cross peaks, shown in structure 1, are described in the text

exchange between protons contributing to broad and narrow resonances can occur in a time of order T_1. In most fluid membrane systems, the rate of lateral diffusion of lipids is rapid[40]. Thus, it is generally assumed that any lipids whose orientation at the magic angle would lead to unexpectedly narrow ^1H NMR resonances should exchange with others making different angles with the applied magnetic field in a time short compared to T_1 (T_1 is often in the range 0.2–1.0 s). This would lead to very different values for the selective and non-selective T_1 measurements. Thus, the magic angle source of the narrow ^1H resonances is highly unlikely. It is much more probable that the molecules giving rise to the narrow lines are compartmentalised in a special structure of some kind, in which they can tumble isotropically to produce resonances distinct from those of the remaining membrane lipid.

IV. Neutral Lipids are Responsible for the NMR Signal

1) Chemical Analysis. Analyses of the enriched plasma membranes of cancer cells[46], and of other types of rapidly dividing cells (such as fibroblasts) known to give narrow NMR resonances, consistently show a high proportion of neutral lipid[47]. Whilst the membrane triglyceride content and composition are comparable in most rapidly dividing cells, levels of other neutral lipid such as cholesterol ester, diglycerides, free cholesterol and other unusual lipids (e.g. o-alkyl ether glycerides) display greater variability.

2) Two Dimensional NMR. Two dimensional scalar correlated spectroscopy (COSY) provides a great deal of information not available from the one dimensional spectrum[63, 64]. Resonances coupled by spin-spin (scalar) mechanisms can be readily identified.

Suspensions of intact viable cancer cells have been studied by two dimensional (2D) NMR methods using a modified pulse sequence[65] (Fig. 10B). The 400 MHz COSY spectrum of triolein in $CDCl_3$ is compared with that obtained from a suspension of cells in Fig. 10. The off diagonal cross peaks, labelled A–G in Fig. 10A indicate a spin coupling between protons on adjacent carbon atoms. The connectivities corresponding to cross peaks A–F, summarised in structure 1, can be seen in the cell spectrum (Fig. 10B), providing unequivocal assignment of lipid acyl chain resonances in the spectrum of cancer cell plasma membranes.

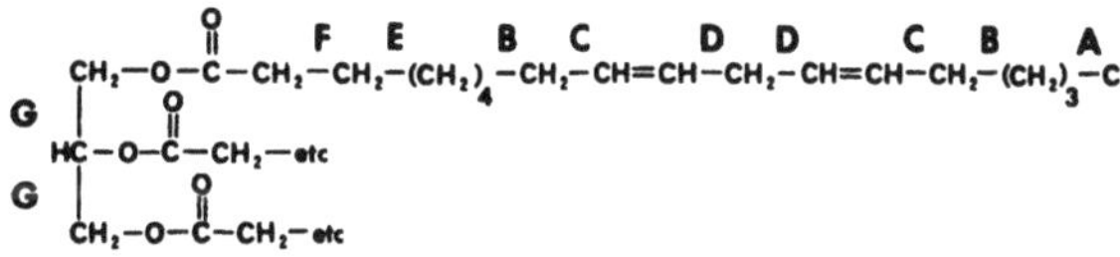

Two dimensional NMR also identifies triglyceride molecules by the presence of a cross peak at 4.3 ppm from the hydroxymethyl glycerol protons designated G′ in structure 1. This resonance, which is unique to triglyceride, is 0.1 ppm downfield from the corresponding glycerol resonance in the phosphatidyl choline spectrum. No cross peaks from a glycerol methylene close to a phosphatidyl choline group are in evidence in the cell spectrum.

In order to determine which cross peaks in the cell spectrum originate from cytoplasmic components such as metabolites, cells were lysed and all the membranes and nuclei spun down, leaving only cytoplasmic components in the supernatant. The COSY spectrum of such a supernatant is shown in Fig. 10C for a comparison with the cell spectrum prior to lysis (Fig. 10B). This supernatant spectrum accounts for most of the remaining cross peaks not generated by lipid molecules.

The amounts of free cholesterol and cholesterol ester vary in the plasma membrane according to the cell type. A cross peak connecting resonances at 0.9 and 1.4 ppm from the methine to methyl coupling in the cholesterol side chain can be observed in the spectra of many cells indicating the presence of one or the other in the lipid domain (Fig. 11B).

3) Comparison with Human Very Low Density Lipoprotein. Lipoproteins are known to contain triglyceride as well as free and esterified cholesterol[67], and as such

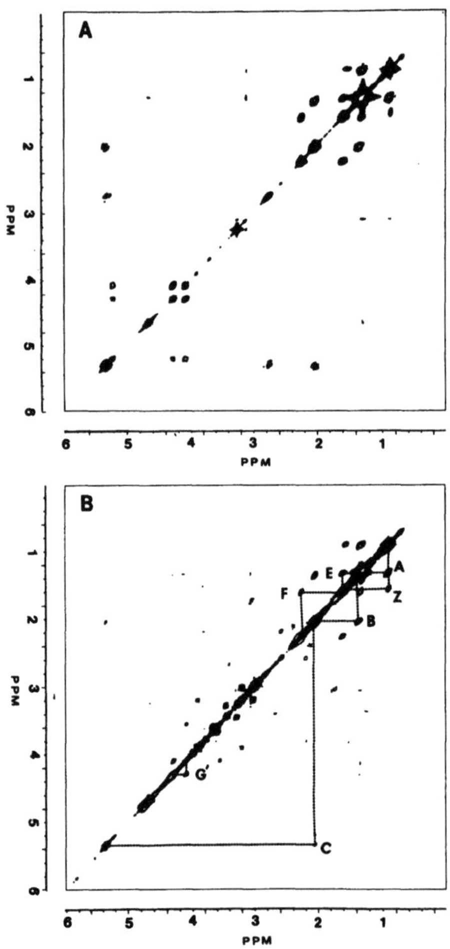

Fig. 11. 400 MHz symmetrised COSY spectra collected at 37°, as previously described, of: – **(A)** Human very low density lipoprotein (VLDL) from healthy donors, suspended in PBS/D_2O. **(B)** Rat mammary adenocarcinoma non-metastatic cell line J clone (1×10^8/400 µl, PBS/D_2O). Lipid acyl chain and glycerol backbone connectivities are indicated. Z denotes the cross peaks between the methyl and methine protons of the alkyl side chain of the cholesterol ring system

are a suitable model with which to compare the triglyceride rich cancer cell plasma membrane.

Human serum lipoproteins were separated into their four main classes: – chylomicra, very low density lipoprotein (VLDL), low density lipoprotein (LDL), and high density lipoprotein (HDL) and their COSY spectra were compared with those obtained from cancer cells[68] (Fig. 11).

The COSY spectrum of VLDL is most similar to that of the cells. The resonances attributable to triglyceride and cholesterol are all present in the cells, except for that of the methine proton, designated G, in the spectrum of triglyceride in $CDCl_3$. Notably absent in the VLDL spectrum are the cross peaks identified as cell supernatant (Fig. 11A).

Selective and non-selective T_1 experiments on chylomicra and VLDL give results similar to those obtained for the cells. Neither LDL nor HDL have the same values for selective and non-selective T_1's, indicating that exchange on the NMR time scale exists between the lipids in these lipoproteins. However, VLDL and chylomicra have within experimental error the same T_1 obtained by selective and non-selective methods (0.4 and 0.6 s, respectively). These data indicate that of the four lipoprotein classes VLDL is most comparable to the plasma membrane bound neutral lipid[68].

4) Triolein in DMSO. The cross peaks in the cell spectrum attributable to triglyceride are noticeably elongated when compared to those from molecules in the supernatant (Fig. 11). A COSY spectrum of triolein in DMSO/H_2O illustrates that the triolein exists in two environments which are seen as separate cross peaks which could account for the elongated cross peaks in the cell spectra. Interestingly the methine-methylene cross peak designated G in the triolein/$CDCl_3$ COSY spectrum is absent for this solvent mixture, suggesting that in the cell membrane the triolein is in an environment with a polarity between that of water and $CDCl_3$. The importance of this observation is made clear when the T_2 values of triolein in $CDCl_3$ and DMSO are discussed in Section F.

V. Summary

These data indicate the existence of lipid domains in or attached to the plasma membrane of cancer cells which i) tumble isotropically, ii) are not in diffusive exchange with remaining membrane lipid and iii) are made up of neutral lipid such as triglycerides along with free and esterified cholesterol.

Although some differences are apparent, the spectrum of human VLDL compares favourably with that of the cells.

E. Biological Considerations for the ^{1}H NMR Experiment

Inherent in the study of living tissue is the difficulty of knowing which variables are a normal part of cell growth and interaction with the environment, and of recognising significant biological differences between the samples in question. As with all biological experiments, the key to reproducible results is to control as many of the variables as possible.

In NMR studies of cancer cells strict control of factors such as growth medium, stage of cell cycle and pH is essential if useful information is to be obtained. Equally important, because of the genetic instability of cancer cells, is a knowledge of the biological status of the cell at the time of the NMR experiment. For example, are cells still as resistant to drugs as they were at the time of the last experiment? Has their capacity to metastasise changed?

This section is an outline of the precautions necessary to minimise the variation in both extrinsic and intrinsic cellular factors, and to ensure that long-term changes in the cells' biology are adequately monitored by techniques other than NMR spectroscopy.

I. Control of Cell Cycle and Growth Medium

Routine tissue culture monitoring procedures should be performed on cells regularly, such as mycoplasma testing, growth curves and effects of different batches of serum. However, in the preparation of samples for NMR experiments additional precautions are necessary. Cells should always be used in a known stage of cell cycle and should be checked by flow cytometric methods[69)].

Cells must have been diluted at least 36 h prior to the experiment to allow time for equilibrium with the changed environment. The medium should be buffered so that there is minimum variation in the pH. Changes in the ^{1}H NMR spectrum similar to those reported in the drug resistant cell lines (Sect. G) can be obtained in the same cell line, simply by varying the pH.

Cells (usually 1×10^8) are spun down and washed four times in PBS/D_2O after removal of medium. Finally the cells are suspended in 400 µl PBS/D_2O and put into a 5 mm NMR tube. Solid tumours are cut up into pieces small enough to fit into a 5 mm NMR tube and washed four times in PBS/D_2O prior to inserting into the tube and covering with 400 µl of the buffer[57)]. During spinning of the sample, in the NMR experiment, the cells spread out and coat the sides of the tube. At the end of the experiment cell viability is measured by trypan blue exclusion; data from samples with a viability less than 85% are not used. Cell lines vary significantly in the time they can remain viable under such conditions, ranging from 3–4 h.

For measurements of the T_2 relaxation parameter, which will be discussed in Sect. F, a settling-down period of one hour after sample preparation is allowed before the experiment is started. With all cancer cell lines investigated thus far, we have observed an increase in the intensity of the sharp methylene resonance at 1.2 ppm, described in Sect. 4, reaching a plateau after approximately one hour.

When changes in the intensity of this resonance are being monitored under different relaxation conditions, such as in the T_2 experiment, it is essential that it should be otherwise stable. The factors contributing to this increase in intensity are not understood, but there are several possibilities. There is much greater cell-cell contact in the NMR tube and some readjustment of the cell membrane is very likely. The pH of the cell suspension drops and plateaus during the first hour as does the ATP level monitored by ^{31}P spectroscopy (Mountford, C. E., Holmes, K. T., Mackinnon, W. B., unpublished data). Whatever the origin of this change in the NMR sensitive domain, the one hour equilibration period is important in order to exclude one source of data variation.

Although considerable time may be necessary to discover which external or cyclical factors are a source of variation and how best to control them, this is a relatively simple task. Much more difficult to monitor and control are the biological changes, both genetic and epigenetic, which occur in all cancer cells with time. Cancer cells are genetically unstable, and over a period of time in culture cells from a single clone will become a heterogeneous population. This poses a major problem in the study of cancer cell lines, particularly in the metastatic and drug resistant models to be discussed in later sections. As in many other laboratories, one of the ways this problem is overcome is to keep large stocks of frozen cells and to culture them for a specified number of divisions before they are discarded. In the case of very unstable cells, thirty to forty doubling times may be all that are allowed, while other lines may be kept much longer. As well as these restrictions, other experiments should be carried out to monitor changes in the cells, and these are outlined below.

II. Flow Cytometry

Many of the cell lines used have chromosome numbers different from the 46 of a normal diploid cell[69]. This should be checked regularly by flow cytometry and once a significant change is detected the particular batch of cells is discarded. For example, in the case of the leukaemic T cell line CCRF-CEM, which is pseudodiploid, a polyploid population appears with time in tissue culture. When this constitutes greater than 10–12% of the population (usually after three months in continuous culture) the cells should no longer be used.

III. Measurement of Drug Resistance

In the case of drug resistant cell lines, which are not always stable[70], a regular measurement of the ID_{50} of the relevant drug should be carried out so that all NMR measurements can be related to the degree of resistance at the time of the experiment.

IV. Animal Experiments

The ability of the cell lines to produce secondary growth (metastases) distant from the site of the primary tumour, must be checked frequently[71]. It appears that ^1H NMR spectroscopy is an exquisitely sensitive technique for monitoring changes in this capacity; however, animal experiments should always be done in parallel as confirmation of the predictive value of the NMR measurements.

It is important that extraction of membranes, animal and NMR experiments, measurement of cellular resistance and ploidy, should all be carried out within a short time span so that complete sets of data are obtained on cells of known biological status. Although such protocols may seem laborious and overly cautious, our experience has been that it is important to measure all these parameters in parallel for each new cellular system.

As with many techniques used for measuring biological systems, NMR is sensitive to fluctuations which occur during the normal course of the cell cycle, thus confusing true biological differences between cell lines. Compounded with this is the natural genetic instability of cancer cells[3, 72, 74]. We have outlined what we consider to be the important sources of variation in experimental data. No doubt, in time, more will become apparent.

F. Cancer Metastasis

NMR spectroscopy has been used extensively to search for differences between benign and malignant tumours. Not until 1984 was the technique applied successfully to cancer metastasis[46, 47, 94].

I. What is Metastasis?

Metastasis is defined by Webster as a shifting of disease from one part of the body to another. This process, thought to be a sequence of inter-related steps, allows cells to escape from the primary site, move unheeded around the body and then lodge and multiply in another site[75, 76].

At the time of diagnosis most malignant neoplasms are biologically heterogeneous and metastases have already occurred in approximately 50% of all cancer patients[77]. The cells in a malignant primary tumour do not have identical metastatic properties. Not only do primary tumours and metastases differ, but metastases can also differ from each other[78]. Tumour heterogeneity poses a major difficulty not only in therapy, but also in basic research and development.

There is now considerable evidence that metastases arise from the outgrowth of a minor subpopulation of tumour cells[78]. Foulds describes the phenomenon of tumour evolution as "neoplastic progression" defining it as "the acquisition of permanent irreversible qualitative changes of one or more characteristics of neoplasm"[79]. However, data from many laboratories now indicate that genetic mutations[42] are only part of the cause. Epigenetic factors can also induce transient changes in the genes[78] and environmental changes may alter membrane structure and function.

To date there is no established method for a rapid evaluation of the metastatic potential of a tumour.

II. Laboratory Metastasis Models

Maintenance of malignant cell lines with unaltered biological characteristics is difficult in the laboratory. To ensure that the cells studied have the same characteristics as the frozen stocks, routine analysis must be undertaken in parallel with any developmental research program. Even during one or two doublings of the cells a major biological alteration such as a change in metastatic capacity can occur. Research into cancer metastasis can therefore only be undertaken effectively with a group of interdisciplinary scientists, all of whom undertake experiments on the *same* batch of cells or excised tumour.

In this laboratory one cell sample is used for injection into rats (for conventional lung colony assays), NMR, isolation and purification of plasma membranes for lipid analysis, cell surface glycoprotein identification and electron microscopy. This requires approximately 2.5×10^9 cells.

1) Rat Metastasis Model. The rat mammary adenocarcinoma malignant metastatic cell line 13762, and its malignant but non-metastatic counterpart J clone[80], has been studied extensively by NMR techniques and its plasma membranes lipids have been analysed.

2) Human Metastasis Models. A human colon carcinoma model developed and supplied by B. Sordat, and human ovarian lines of varying metastatic potential developed and supplied by Carolien van Haaften-Day, have been studied by NMR relaxation methods. However, human metastasis models were found to have rapidly altering biological characteristics which were difficult to control in culture and nude mice.

III. Transverse Relaxation (T_2) is Altered in Metastatic Cell Membranes

Both metastatic 13762 cells and non-metastatic J clone cells gave four clearly resolved resonances under the broad methylene resonance (1.2 ppm) of the ^1H NMR spectrum. Application of the Meiboom-Gill modification of the Carr-Purcell (CPMG) pulse sequence[45] revealed different relaxation rates for these subcomponents; further resolution enhancement could be achieved by varying the interpulse delay. Both single cell suspensions and solid tumours gave the profiles observed for the plot of the unmodified spectra in Fig. 12A. The enhancement technique clearly resolved the resonances underlying the broad methylene resonance for most cell suspensions; however, the line widths from a solid tumour can be much broader, and separation of the underlying resonances was sometimes difficult.

In the resolution-enhanced CPMG spectra of the metastatic line 13762 cells, four resonances at 1.22, 1.23, 1.25 and 1.27 ppm were resolved under the broad acyl chain envelope (1.2 ppm). Each of these four resonances showed two rates of decay (Fig. 12B). In contrast, the unmodified broad $-CH_2-$ signal showed three apparent rates of decay, which were a composite of the three underlying peaks and were attributed to the resonance at 1.22 over the delay times 0–20 msec and 30 to 80 ms. The third and slowest rate was clearly due to the resonance at 1.25 ppm over the range 80 to 150 ms. This distinguished the cells with metastatic potential from the non-metastatic variant (Fig. 12B). This longest T_2 is now denoted T_{2l}.

The resolution-enhanced spectra of the non-metastatic J clone cells also had four resonances at 1.22, 1.23, 1.25 and 1.27 ppm, but it was difficult to determine which,

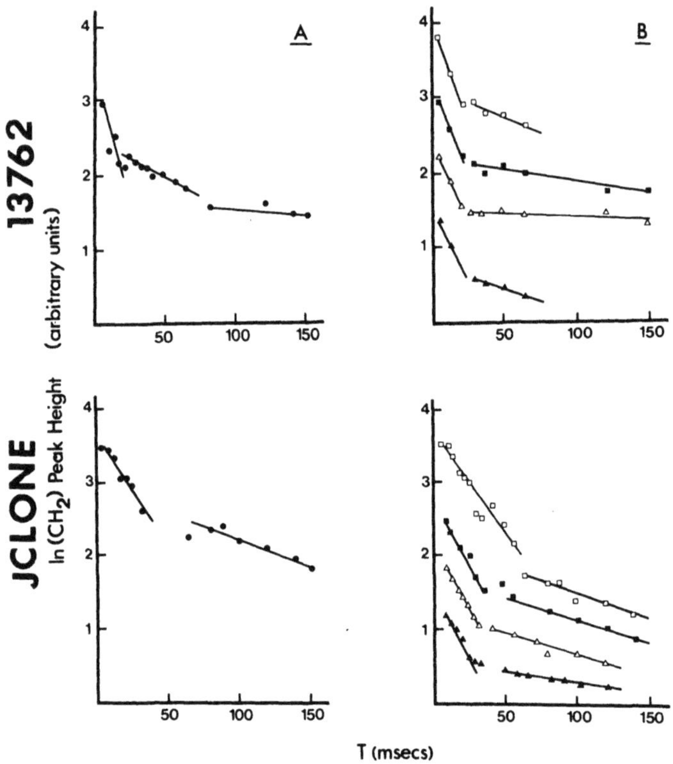

Fig. 12. The Meiboom-Gill modification of the Carr-Purcell pulse sequence was applied to the mammary adenocarcinoma metastatic cell line 13762 and the non-metastatic line J clone. The pulse sequence was excited as described[45]. Thirty-two accumulations were recorded with the water resonance suppressed by gated irradiation[91]. **(A)** Intensity of the methylene resonance plotted against the delay between the first pulse and the nth echo. These spectra had a line-broadening of 3 Hz. Only one methylene resonance was observable. **(B)** Similar to **(A)** but with a Lorentzian-Gaussian resolution enhancement (K = 11 Hz; g = 0.04; a = 1.64 seconds). Four resonances are observable at 1.22 ppm (□), 1.23 ppm (■), 1.25 ppm (△) and 1.27 ppm (▲)

Table III. Transverse relaxation parameter (T_2) for the methylene resonance in 13762 and J clone cells

Chemical shift (ppm)	T_2 (milliseconds) 13762 cells	J clone cells
	Unmodified Spectra	
1.2	21 ± 3	32 ± 6
	86 ± 16	154 ± 18
	772 ± 50	
	Resolution-enhanced spectra	
1.22	19 ± 4	50 ± 14
	99 ± 27	136 ± 42
1.23	42 ± 12	63 ± 17
	167 ± 64	224 ± 44
1.25	28 ± 3	55 ± 13
	797 ± 106	117 ± 20
1.27	31 ± 4	38 ± 9
	162 ± 52	157 ± 36

Values for T_2 were calculated by a least-squares method. All values gave an r^2 greater than 0.98. Results are expressed as the mean ± standard error for three experiments

if any, dominated the two rates observed in the unmodified $-CH_2-$ resonance (Fig. 12A). However, the resonance at 1.25 ppm did not have the long T_2 observed in the metastatic parent line (see Table III).

A phenotypic drift of the J clone toward the metastatic parent line was observed twice in a solid tumour grown from cells that were injected subcutaneously into Fisher rats and were originally believed to lack metastatic potential[46)]. The results of CPMG experiments were characteristic of neither cell line 13762 nor J clone, but of a composite of the two. The lungs of all tumour-bearing animals were sectioned and counted[81)]. The partially reverted J clone tumour (in a rat inoculated with 10^8 cells) showed a mean of nine colonies in each 5 μm section, each colony containing 100 to 125 cells. A non-reverted J clone tumour showed no metastases at all, whereas the 13762 tumour (in a rat inoculated with 10^7 cells) showed the highest incidence of metastatic replacement, with 85 percent of the lung tissue involved.

The human colon metastasis model showed the presence of long T_2's of 400 ms or above, in the cell line known to have metastatic potential.

Human ovarian lines with varying degrees of metastatic potential also showed the presence of long T_2's for the resonance at 1.25 ppm.

IV. How is Such a Long T_2 Generated?

The T_{21} of greater than 300 ms was found to correlate with the presence of malignant cells capable of metastasis. T_{21} values are usually long and can only be accounted for by fast molecular motion in an unusual environment. We have shown earlier that triglycerides are responsible for at least part of this narrow resonance. Different solvent systems have been studied in an attempt to determine the environment conducive to this type of relaxation mechanism[47].

Both triglycerides and cholesterol esters of differing acyl chain content, in $CDCl_3$, were found to give single rates of decay with T_2 values varying from 1 s to 300 ms. The acyl chains from cholesterol esters and triglycerides were found to behave identically. Neutral lipids with one double bond at position 9 generate two clusters of resonances at 1.28 and 1.30 ppm from the nuclei at positions 2–7 and 12–17, in the chain respectively[82]. The results of a CPMG experiment, over the delay range 0–2 s for triolein (18:1) is shown in Fig. 13A.

In contrast, in $DMSO/H_2O$ (99.8%:0.2%) triolein generates four resonances at 1.25, 1.27, 1.30 and 1.35 ppm which suggests that the lipid exists in two environments. A CPMG experiment over the delay range 0–2 s generates three apparent T_2 values for the resonance at 1.25 ppm of 620, 940 and 1600 ms; the resonance at 1.35 ppm has two T_2 values of 500 and 740 ms. Resonances at 1.27 and 1.30 ppm have single T_2 values of 253 and 142 ms respectively (Fig. 13B). Triolein dispersed

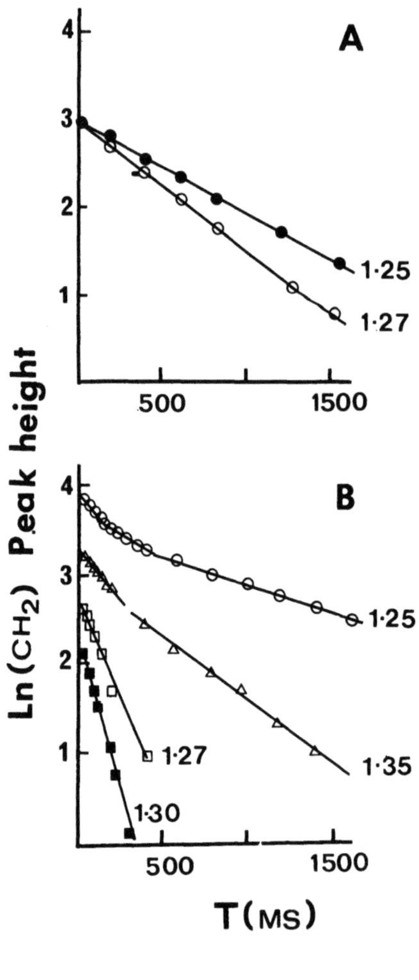

Fig. 13. The natural logarithm of the methylene resonance intensity is plotted against the delay between pulses in the CPMG experiment. **(A)** Triolein (20 mg/ml) in $CDCl_3$; **(B)** Triolein 2 mg/ml in DMSO : H_2O (99.8% : 0.2%). Chemical shifts of the resonances are given on each plot

in D_2O in a variety of ways does not yield a T_2 in excess of 200 ms. Hence, triolein in a solvent with a polarity between water and chloroform has relaxation behaviour similar to that of the neutral lipids in the living cell membrane.

V. Clinical Studies

Human colon and ovary samples removed during surgery have been examined by NMR relaxation methods[47)]. Data obtained from primary colon and ovarian cancers, shown in Fig. 14, can be seen to be comparable with those obtained for the rat mammary adenocarcinoma metastasis model (Fig. 12).

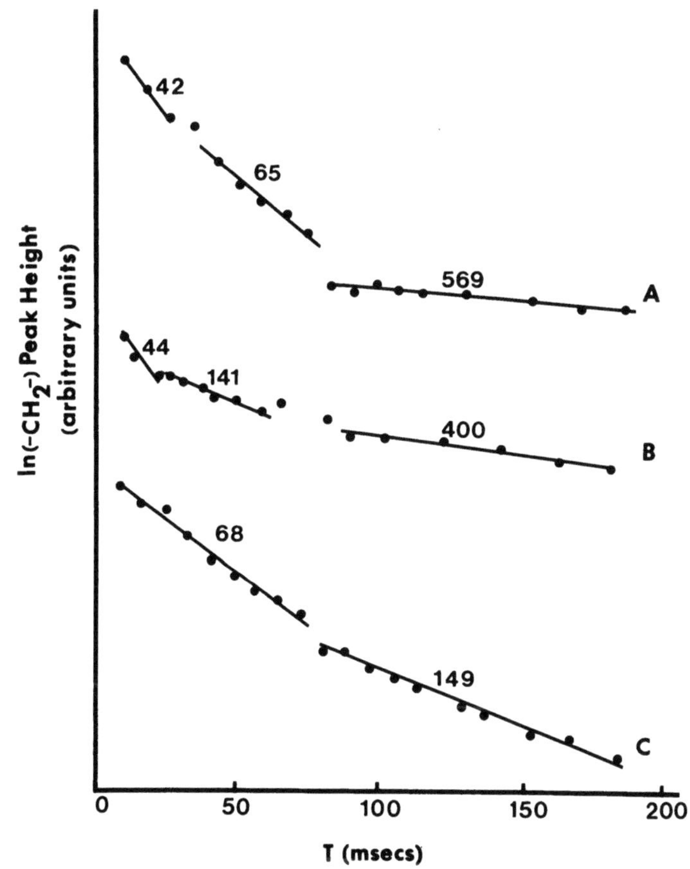

Fig. 14. The Meiboom-Gill modification of the Carr-Purcell pulse sequence was applied to human tumours. The sample was prepared and the pulse sequence executed as described previously[45)]. Thirty-two accumulations were recorded with the water resonance suppressed by gated irradiation[91)]. The intensity of the $-CH_2-$ resonance is plotted against the delay between the first pulse and the nth echo. The spectra had an applied line broadening of 3 Hz. **(A)** Primary ovarian tumour (data obtained at 300 MHz). **(B)** Secondary peritoneal tumour from either a gastrointestinal or ovarian primary source (data obtained at 400 MHz). **(C)** Primary colon tumour (data obtained at 300 MHz). The T_2 (ms) (indicated by each slope) were calculated by a least squares fit where $R^2 \geq 0.98$

Most of the biopsy samples examined have been suitable for study by ^1H NMR. The T_2 relaxation data obtained from 32 of the 34 human tumour samples examined indicate that this method is suitable for use on excised human tumours. However, unlike the laboratory models where conventional lung colony assays on rats rapidly established the tumour cell metastatic capability, the NMR spectra obtained from human biopsy samples await report of the patients' progress for evaluation.

Compared to established histological methods' for determining malignancy, NMR is fast and precise, providing data which can be interpreted within three hours of biopsy. If a correlation exists between the patient's progress and NMR relaxation parameters this method could be of prognostic use and eliminate post operative adjuvant chemotherapy for those patients with non-metastatic primary tumours.

G. Drug Resistance

I. Drug Resistance

The development of cellular resistance to anti-cancer drugs is a major problem in the control of the disease. After treatment some surviving cells are able to multiply in the presence of a drug, which on prior exposure was toxic to the large bulk of the tumour population. Many experimental models of cellular drug resistance have been developed and some of the *in vitro* mechanisms of resistance have been demonstrated clinically[77, 84)].

Pleiotropic drug resistance, whereby cells are resistant to a range of drugs unrelated to the initial agent, is accompanied by changes in the plasma membrane. These include an increase in the amount of 170,000 Dalton glycoprotein in the plasma membrane of resistant cells (P-glycoprotein)[85, 86)], as well as changes in the lipid structural order[87, 88)]. The mechanism of resistance is still undetermined but the net effect is to reduce uptake of drug by the cell and thereby render it less sensitive.

II. NMR Spectroscopy of Resistant Cells

The human leukaemic T cell line CCRF-CEM can be made resistant to the anti-cancer drug vinblastine by culturing in an increasing sub-lethal concentration of the drug. Figure 15 shows the spectra of the sensitive parent line compared with lines resistant to 10 and 20 ng/ml vinblastine, called VBL10 and VBL20 respectively. The most notable change is an increase in the intensity of the $-CH_2-$ resonance at 1.2 ppm upon the development of drug resistance. Kinetic studies, in which vinblastine is added directly to the cells in the NMR tube[48)], indicate a difference between resistant and sensitive cells in the behaviour of this resonance on exposure to the drug. Using resolution enhancement, changes in intensity of the resonance at 1.25 ppm can be followed for an hour (Fig. 16). This peak, generally the most intense and best resolved, is affected by vinblastine differently in the resistant cells

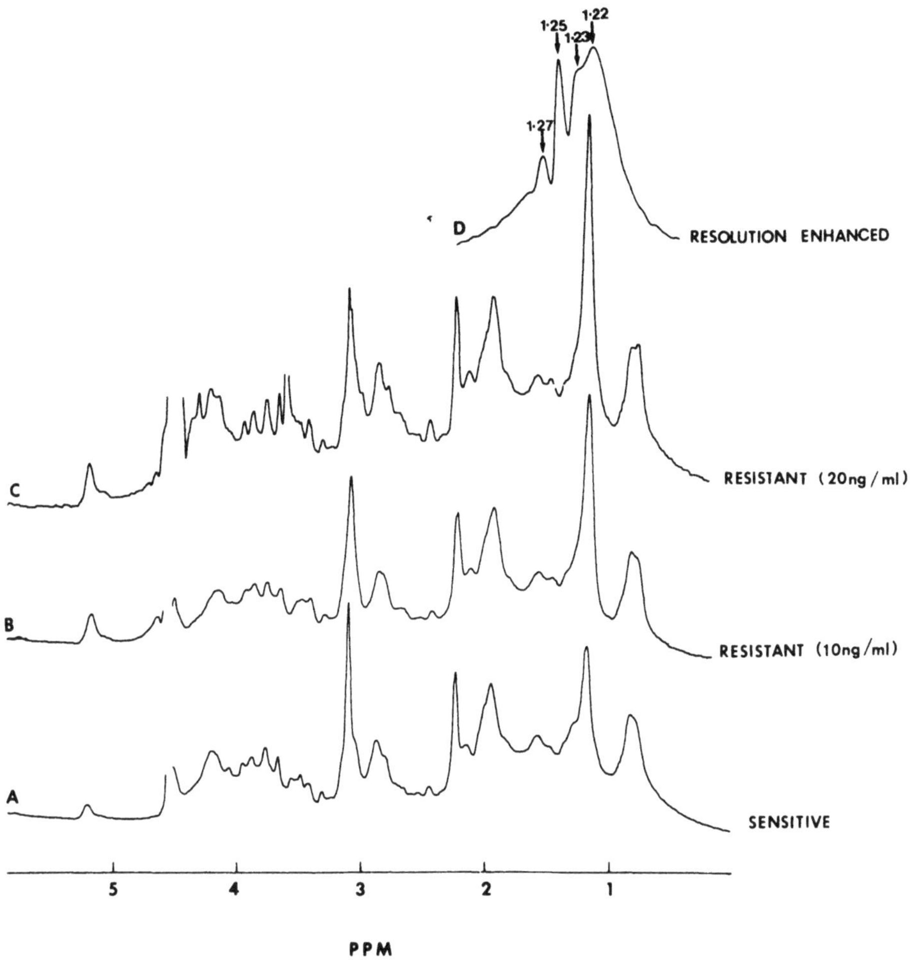

Fig. 15. Comparison of the 400 MHz ^{1}H NMR spectrum of CCRF-CEM lines resistant to (**A**) 20 ng/ml vinblastine, and (**B**) 10 ng/ml vinblastine with (**C**) the sensitive parent line. Spectra were collected at 37° on a Bruker WM 400 MHz spectrometer with the residual water suppressed by selective gated irradiation[91)]

compared to the sensitive line. At 40 ng/ml of vinblastine there is a marked decrease in intensity relative to resistant cells with no drug, whereas there is little difference in the behaviour of this peak in sensitive cells with or without drug. These data are difficult to interpret but indicate that the NMR sensitive domain has changed with the development of drug resistance.

The two dimensional COSY technique has also been used to study drug resistance[48)]. The 400 MHz spectrum of the cells resistant to 20 ng/ml vinblastine (VBL20) is compared with the sensitive line in Fig. 17A and B, respectively. A major difference between the two is a cross peak between 2.2 and 3.8 ppm in the resistant line which is absent in the sensitive cells. This is part of the supernatant

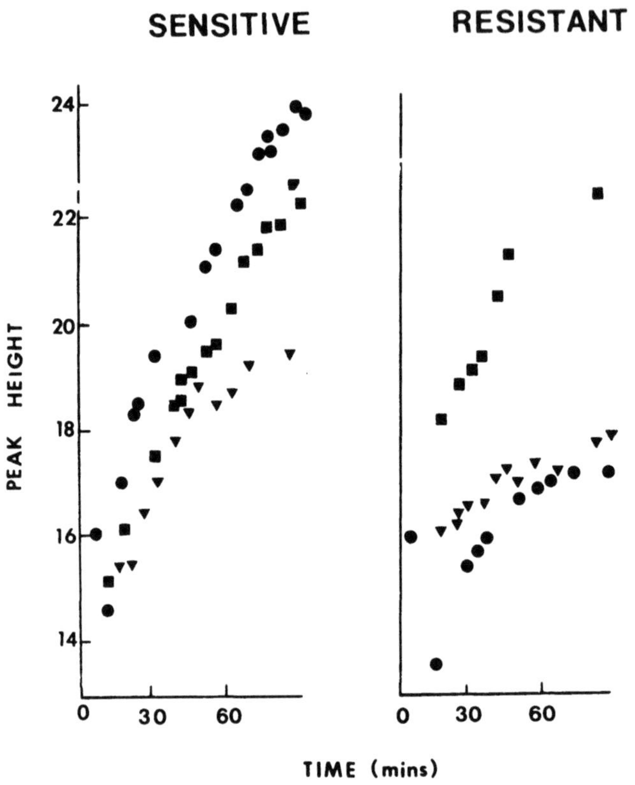

Fig. 16. Kinetic effect of vinblastine on the methylene resonance at 1.25 ppm, obtained with Lorentzian-Gaussian resolution enhancement (K = 11 Hz, g = 0.04; a = 1.64 seconds). Vinblastine is added to the NMR tube at 0 min at a concentration of 20 ng (■) and 40 ng (●)/ml. The control sample (▼) has no drug. Sensitive CCRF-CEM cells were compared with cells resistant to 20 ng/ml vinblastine

component of the COSY spectrum discussed in Sect. D (IV) and is consistent with a cross peak from sodium glutamate, indicating an increase in the concentration of this metabolite in drug-resistant cells. In general, the cross peaks in the sensitive CCRF-CEM line spectrum are weaker than in the VBL20 spectrum particularly those from cytoplasmic components. This is consistent with the general increases in lipid and protein content of the VBL20 cells and their enhanced respiratory function (Wright, L. C., Dyne, M., Holmes, K. T., Romeo, A., Mountford, C. E., submitted for publication.

NMR spectroscopy can thus be used to study changes in the cell membrane upon the development of drug resistance, as well as alterations in cell metabolism.

III. Low Level Drug Resistance

Most reports of cellular resistance to anti-cancer drugs are on cells with between 200 and 2000-fold resistant[90]. The most common method of developing these is by continuous exposure to increasing sublethal doses of the drug. However, neither the high level of resistance, nor the method of generating it may be clinically significant.

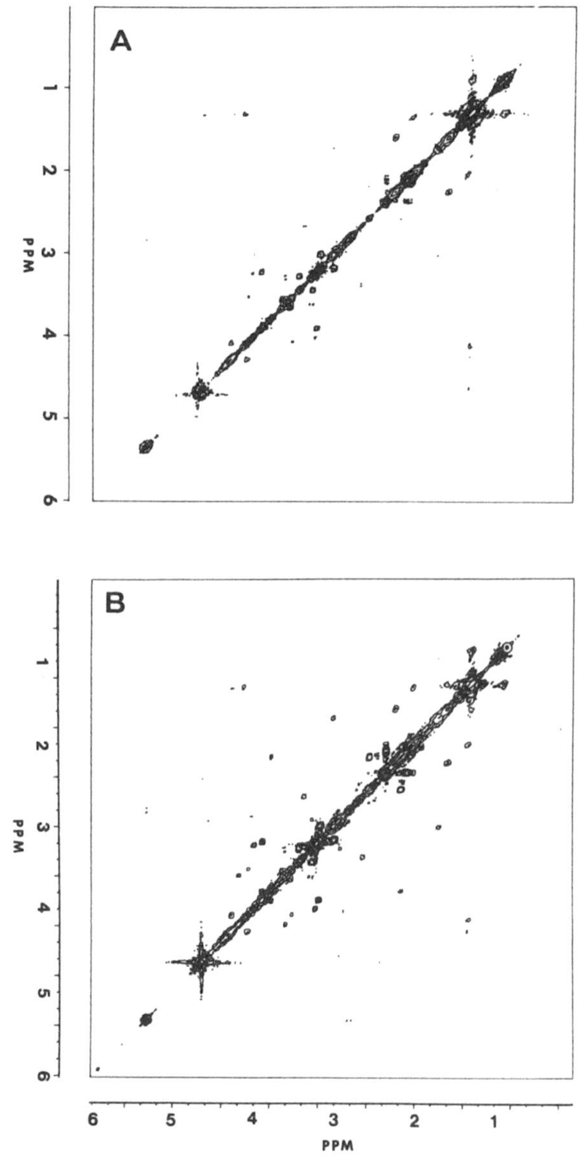

Fig. 17. 400 MHz symmetrised COSY spectra collected at 37°, as previously described[65]. **(A)** Sensitive CCRF-CEM cells. **(B)** CCRF-CEM cells resistant to 20 ng/ml vinblastine (VBL 20)

The 20-fold resistance to VBL developed for study by ^{1}H NMR is low and unstable. Over a period of weeks, it spontaneously reverts to an even lower level of resistance. This type of drug resistance, which is perhaps closer to the clinical situation, is temporary and hence any genetic alteration which may occur is reversible. Nevertheless, the plasma membrane alterations recorded by ^{1}H NMR are consistent with reduced permeability to the drug and the membrane domains in question may even be involved in the drug transport. Cellular uptake of cholesterol ester is via lipoproteins, and recently LDL has been used to facilitate the entry of daunorubicin into cancer cells[95]. It is conceivable that the passage of vinblastine, a cholesterol like molecule, across the plasma membrane could be affected by changes in the triglyceride-rich lipid domain detected by ^{1}H NMR.

H. Magnetic Resonance Imaging – What is Really Being Measured?

Magnetic resonance imaging spectrometers make use of the total intensity of the proton NMR spectrum. Usually no Fourier Analysis is done to separate the intensity into components from various chemical species. If a spin-echo acquire sequence is used, the time between the echo-forming pulses can be adjusted so as to minimise contributions from large molecules with short T_2 values. As such it has been assumed that in MRI one is only using the water resonance. In healthy cells this assumption would normally be correct since water is such a major component and many of the large molecules (DNA, protein, membrane lipid) will have very wide resonances. Nevertheless, all proton-containing components will contribute to the first data points on the free induction decay acquired by the spectrometer and as we have seen in this article, some cells contain sufficient lipid in effectively isotropic structures to yield a high resolution ^{1}H NMR spectrum with a significant contribution from the methylene groups of fatty acyl chains. These narrow resonances have relatively long T_2 values and hence will refocus in the echo sequence along with the water component. Thus, MRI images may comprise lipid as well as water, and the various contrast routines using either T_1 or T_2 sequences may enhance or diminish contributions from these species. These non-specific relaxation times may confuse the issue in correlating MRI with particular pathological conditions.

The techniques of spatial localisation with chemical shift determination[26, 27] can yield estimates of the populations of components contributing to the MRI intensity. This procedure would thus be highly desirable in any attempts to correlate pathological state with relaxation times.

J. NMR in Cancer – The Prognosis

The present state of NMR technology is more than adequate for a large number of accurate measurements. The precision with which NMR reported the unique chemical content and rapid diversification of cancer cells has generated confusion in the past. Now with tumour heterogeneity becoming better understood NMR will undoubtedly contribute towards a better understanding of cancer. The necessary caveats have been expressed, and it is expected that the three aspects of NMR in cancer research, imaging, metabolism, and membrane structure will provide much practical as well as theoretical information about the malignant state. With the very wide availability of NMR instruments in hospitals and biology departments, one can expect that this progress will take place very rapidly.

K. Acknowledgements

We would like to acknowledge Professors Antony Basten, Myer Bloom, Richard M. Fox, John K. Saunders, Peter E. Wright and current and past members of this research team for their scientific contribution and dedication. Special thanks goes to Professor Martin Tattersall for his suggestion to investigate the use of NMR spectroscopy in tumour heterogeneity and his confidence in our ability to develop this technology. We are indebted to Drs. Ian Ramshaw, Bernard Sordat and Carolien van Hafften-Day for supplying us with their metastasis models and maintaining an active interest in our progress. Finally we thank Judy Hood for her enthusiastic typing of the manuscript.

L. Abbreviations

ATP =	Adenosine triphosphate	LDL =	Low density lipoprotein
$CDCl_3$ =	Deuterated chloroform	MHz =	Megahertz
COSY =	Two dimensional scalar correlated spectroscopy	MRI =	Magnetic resonance imaging
CPMG =	Meiboom-Gill modification of the Carr-Purcell pulse sequence	NMR =	Nuclear magnetic resonance
		PBS =	Phosphate buffered saline
D_2O =	Heavy water	PCr =	Phosphocreatine
DMSO =	Dimethyl sulphoxide	P_i =	Inorganic phosphate
FU =	Fluorouracil	PME =	Phosphomonoester
HDL =	High density lipoprotein	ppm =	Parts per million
ID_{50} =	Concentration of drug which reduces the cellular growth to half that of the untreated cells over a defined time	T_1 =	Longitudinal relaxation
		T_2 =	Transverse relaxation
		VBL =	Vinblastine
		VLDL =	Very low density lipoprotein

M. References

1. Damadian, R.: Science *171,* 1151 (1971)
2. Bottomley, P. A., Hardy, C. J., Argersinger, R. E., Allen, G. R.: in preparation, see Abstracts for Soc. Mag. Res. in Medicine *1,* 29 Aug. 19–23 (1985)
3. Schirrmacher, V.: Adv. Cancer Res. *43,* 1 (1985)
4. Abragam, A.: Principles of Nuclear Magnetism. Clarendon Press, Oxford 1961
5. Farrar, T. C., Becker, E. D.: Pulse and Fourier Transform NMR, New York, Academic Press 1971
6. Becker, E. D.: High Resolution NMR, New York, Academic Press 1980

7. Brindel, K. M., Brown, F. F., Campbell, I. D., Grathwohl, C., Kuchel, P. W.: Biochem. J. *180,* 37 (1979)
8. Freeman, R., Hill, H. D. W.: J. Chem. Phys. *51,* 3140 (1969)
9. Hore, P. J.: J. Magn. Res. *55,* 283 (1983)
10. Benz, F. W., Feeney, J., Roberts, G. C. K.: J. Magn. Res. *17,* 128 (1975)
11. Smith, I. C. P., Ekiel, I. H.: Phosphorus-31 NMR of phospholipids in membranes, in: Phosphorus-31 NMR, Principles and Applications (ed.) Gorenstein, D. G., p. 447, Orlando, Florida, Academic Press 1984
12. Bloom, M., Smith, I. C. P.: Manifestations of lipid-protein interactions in deuterium NMR, in: Progress in Protein-Lipid Interactions, Vol. 2 (eds.) Watts, A. and De Pont, J. J. H. H. M., p. 61, Amsterdam, North Holland 1985
13. Smith, I. C. P.: NMR of liquid crystalline lipids in biological membranes, in: Nuclear Magnetic Resonance of Liquid Crystals (ed.) Emsley, J. W., p. 533, Dordrecht, Reidel 1985
14. Davis, J. H.: Biochim. Biophys. Acta *737,* 117 (1983)
15. Wennerström, H., Lindblom, G.: Q. Rev. Biophys. *10,* 67 (1977)
16. MacKay, A. L.: Biophys. J. *35,* 301 (1981)
17. Mansfield, P., Morris, P. G.: NMR Imaging in Biomedicine, New York, Academic Press 1982
18. Young, S. W.: Nuclear Magnetic Resonance Imaging. Basic Principles, New York, Raven Press 1984
19. Hopf, M.-A., Smith, F. W. (eds.): Magnetic Resonance in Medicine and Biology, Basel, Karger 1984
20. Ackerman, J. J. H., Grove, T. H., Wong, G. C., Gadian, D. G., Radda, G. K.: Nature *283,* 167 (1980)
21. Griffiths, J. R., Stevens, A. N., Iles, R. A., Gordon, R. E., Shaw, D.: Bioscience Reports *1,* 319 (1981)
22. Deslauriers, R., Geoffrion, Y., Butler, K. W., Smith, I. C. P.: Quart. Rev. Biophys. *18,* 65 (1985)
23. Ross, B. D., Radda, G. K., Gadian, D. G., Rocker, G., Isiri, M., Falconer-Smith, J.: New England J. Med. *304,* 1338 (1981)
24. Wilson, J. R., Fink, L., Maris, J., Ferraro, N., Power-Vanwart, J., Eleff, S., Chance, B.: Circulation *71,* 57 (1985)
25. Hope, P. L., Costello, A. M., de L., Cody, E. B., Delpy, D. T., Tofts, P. S., Chu, A., Hamilton, P. A., Reynolds, E. P. R., Wilkie, D. R.: Lancet 366 (1984)
26. Aue, W. P., Müller, S., Seelig, J.: J. Magn. Res. *61,* 392 (1985)
27. Bottomley, P. A., Edelstein, W. A., Foster, T. H., Adams, W. A.: Proc. Natl. Acad. Sci. U.S.A. *82,* 2148 (1985)
28. Agris, P. F., Campbell, I. D.: Science *216,* 1325 (1982)
29. Haselgrove, J. C., Eleff, S., Leigh, J. S., Gyulai, L., Bolinger, L., Subramian, H. V., Chance, B.: Continuous non-invasive organ biochemistry and NMR imaging of brain, in: Brain Imaging and Brain Function (ed.) Sokoloff, L., p. 271, New York, Raven Press 1985
30. Griffiths, J. R., Cady, E., Edwards, R. H. T., McCready, V. R., Wilkie, D. R., Wiltshaw, E.: Lancet 1435 (1983)
31. Ng, T. C., Evanochko, W. T., Hiramoto, R. N., Ghanta, V. K., Lilly, M. B., Lawson, A. J., Corbett, T. H., Durant, J. R., Glickson, J. D.: J. Magn. Res. *49,* 271 (1982)
32. Evanochko, W. T., Ng, T. C., Lilly, M. B., Lawson, A. J., Corbett, T. H., Durant, J. R., Glickson, J. D.: Proc. Natl. Acad. Sci. U.S.A. *80,* 334 (1983)
33. Evanochko, W. T., Sakai, T. T., Ng, T. C., Krishna, N. R., Kim, H. D., Zeilder, R. B., Ghanta, V. K., Brockman, R. W., Schiffer, L. M., Brownschweiger, P. G., Glickson, J. D.: Biochim. Biophys. Acta *805,* 104 (1984)
34. Lilly, M. B., Ng, T. C., Evanochko, W. T., Katholi, C. R., Kumar, N. G., Elgavish, G. A., Durant, R. A., Hiramoto, R., Ghanta, V., Glickson, J. D.: Cancer Res. *44,* 633 (1984)
35. Prysor-Jones, R. A., Silverlight, J. J., Jenkins, J. S., Stevens, A. N., Rodrigues, J. L., Griffiths, J. R.: FEBS Letts. *177,* 71 (1984)
36. Maris, J. S., Evans, A. E., McLaughlin, A. C., D'Angio, G. J., Bolinger, L., Manos, H., Chance, B.: New England J. Med. *312,* 1500 (1985)
37. Stevens, A. N., Morris, P. G., Iles, R. A., Sheldon, P. W., Griffiths, J. R.: Br. J. Cancer *50,* 113 (1984)

38. Malet-Martino, M.-C., Martino, R., Lopez, A., Beteille, J.-P., Bon, M., Bernadou, J., Armand, J.-P.: Cancer Chemother. Pharmacol. *13,* 31 (1984)
39. Bernadou, J., Armand, J. P., Lopez, A., Malet-Martino, M. C., Martino, R.: Clinical Chemistry *31,* 846 (1985)
40. Lindblom, G., Johansson, B.-O., Arvidson, G.: Biochemistry *20,* 2204 (1981)
41. Bottomley, P. A., Foster, T. H., Argersinger, R. E., Pfeifer, L. M.: Med. Phys. *11,* 425 (1984)
42. Nowell, P.: In Tumour heterogeneity. Origins and Implications (ed.) Owens, A. M., Coffey, D. S., Baylin, S. B., p. 22: 351, Academic Press, New York 1982
43. Damadian, R., Zaner, K., Hor, D., DiMaio, A., Minkoff, L., Goldsmith, M.: Ann. N.Y. Acad. Sci. *222,* 1048 (1973)
44. Inch, W. R., McCredie, J. A., Knispel, R. R., Thompson, R. T., Pintar, M. M.: J. Nat. Cancer Inst. *52,* 353 (1974)
45. Mountford, C. E., Mackinnon, W. B., Burnell, E. E., Bloom, M., Smith, I. C. P.: J. Biochem. Biophys. Meth. *9,* 323 (1984)
46. Mountford, C. E., Wright, L. C., Holmes, K. T., Mackinnon, W. B., Gregory, P., Fox, R. M.: Science *226,* 1415 (1984)
47. May, G. L., Wright, L. C., Holmes, K. T., Williams, P. G., Smith, I. C. P., Wright, P. E., Fox, R. M., Mountford, C. E.: J. Biol. Chem., in press (1986)
48. Holmes, K. T., Fox, R. M., Wright, L. C., Mountford, C. E.: Membrane alterations on the development of drug resistance in human lymphocytes, in: NMR in Cancer (ed.) Allen, P., Pergamon Press (in press) 1986
49. Block, R. E., Maxwell, G. P., Prudhomme, D. L., Hudson, J. L.: J. Natl. Cancer Inst. *58,* 151 (1977)
50. Block, R. E.: FEBS Lett. *34,* 109 (1973)
51. Block, R. E., Maxwell, G. P., Irvin, G. L., Hudson, J. L., Prudhomme, D. L.: Nuclear magnetic resonance studies of membranes of normal and cancer cells, in: Membrane transformations in neoplasia (eds.) Schultz, H., Block, R. E., p. 244, New York, Academic Press 1974
52. Block, R. E.: Proton magnetic resonance characteristics of tissues: High resolution studies of water and lipid signals from normal and tumor rat tissues, in: The Nuclear Resonance Effect in Cancer (ed.) Damadian, R., Chapter 10, Portland, Oreg., Pacific Publishing Company 1976
53. Nicolau, C., Dietrich, W., Steiner, M. R., Steiner, S., Melnick, J. L.: Biochim. Biophys. Acta *382,* 311 (1975)
54. Nicolau, C., Klenk, H.-D., Reimann, A., Hilderbrand, I. C., Bauer, M.: Biochim. Biophys. Acta *511,* 83 (1978)
55. Mountford, C. E., Grossman, G., Hampson, A. W., Holmes, K. T.: Biochim. Biophys. Acta *720,* 65 (1980)
56. Mountford, C. E., Grossman, G., Gatenby, P. A., Fox, R. M.: Br. J. Cancer *41,* 1000 (1980)
57. Mountford, C. E., Grossman, G., Reid, G., Fox, R. M.: Cancer Res. *42,* 2270 (1982)
58. Bergelson, L. D., Barsukov, L. I.: Science *197,* 224 (1977)
59. Bloom, M., Burnell, E. E., MacKay, A. L., Nichol, C. P., Valic, M. I., Weeks, G.: Biochemistry *17,* 5750 (1978)
60. Ulmius, J., Wennerström, H., Lindblom, G., Arvidson, G.: Biochim. Biophys. Acta *389*, 197 (1975)
61. Wennerström, H., Ulmius, J.: J. Mag. Res. *23,* 431 (1976)
62. Bloom, M., Burnell, E. E., Roeder, S. B. W., Valic, M. I.: J. Chem. Phys. *66,* 3012 (1977)
63. Wagner, G., Kumar, A., Wüthrich, K.: Eur. J. Biochem. *114,* 375 (1981)
64. Arseniev, A. S., Wider, G., Jonbert, F. J., Wüthrich, K.: J. Mol. Biol. *159,* 323 (1982)
65. Cross, K. J., Holmes, K. T., Mountford, C. E., Wright, P. E.: Biochemistry *23,* 5895 (1984)
66. Yushmanov, V. E., Kurushin, E. A., Kogarko, I. N., Sibeldina, L. A., Mokeeva, R. A., Lukina, E. A.: Biological Membranes (USSR) *2,* 170 (1985)
67. Skipski, V. P.: Lipid Composition of lipoproteins in Normal and Diseased States, in: Quantitation and Composition of Metabolism (ed.) Nelson, G. J., *11,* 471, New York, Wiley 1972
68. Williams, P. G., Helmer, M. A., Wright, L. C., Dyne, M., Fox, R. M., Holmes, K. T., May, G. L., Mountford, C. E.: FEBS Lett. *192,* 159 (1985)
69. Barlogie, B., Raber, M. N., Schumann, J., Johnson, T. S., Dreqinko, B., Swartzendruber, D. E., Gohde, W., Andreef, M., Freireich, E. J.: Cancer Res. *43,* 3982 (1983)
70. Beck, W. T., Mueller, T. J., Tanzer, L. R.: Cancer Res. *39,* 2070 (1979)

71. Fidler, I. J., Hart, J. R.: Science *217,* 998 (1982)
72. Fidler, I. J., Gersten, D. M., Hart, I. R.: Adv. Cancer Res. *28,* 149 (1978)
73. Turner, D. A.: Seminars in Nuclear Medicine, *XV,* 210 (1985)
74. Sugarbaker, E. V.: Cancer Metastasis: A product of tumor host interactions, in: Current Problems in Cancer *3,* 3 (1979)
75. Fidler, I. J.: Cancer Res. *45,* 4714 (1985)
76. Nowell, P. C.: Science *194,* 23 (1976)
77. Schimke, R. T.: Cancer Res. *44,* 1735 (1984)
78. Frost, P., Kerbel, R. S.: On a possible epigenetic mechanism of tumor cell heterogeneity the role of DNA methylation. Cancer Metastasis Rev. *2,* 375 (1983)
79. Foulds, L.: J. Natl. Cancer Inst. *17,* 701 (1956)
80. Ramshaw, J. A., Carlsen, S. T., Moon, D., Warrington, R. C.: Int. J. Cancer *30,* 601 (1982)
81. Boeryd, B., Ganeluis, T., Lundin, P., Mellgren, J.: Int. J. Cancer *1,* 497 (1966)
82. Frost, D. J., Gunstone, F. D.: Chem. Physics and Lipids *15,* 53 (1975)
83. Irving, M. G., Simpson, S. J., Field, J., Doddrell, D. M.: Cancer Res. *45,* 481 (1985)
84. Bell, D. R., Gerlach, J. H., Kartner, N., Buick, R. N., Ling, V.: J. Clin. Onc. *3,* 311 (1985)
85. Ling, V., Thompson, L. H.: J. Cell Physiol. *83,* 103 (1974)
86. Naruse, S., Hirakaiva, K., Horikaiva, Y., Tanake, C., Higuchi, T., Veda, S., Nishikawa, H., Watari, H.: Cancer Res. *45,* 2429 (1985)
87. Ramu, A., Glaubieger, D., Macgrath, I., Joshi, A.: Cancer Res. *43,* 5533 (1983)
88. Siegfried, J. A., Kennedy, K. A., Sartorelli, A. C., Tritton, T. R.: J. Biol. Chem. *258,* 339 (1983)
89. Ng, T. C., Majors, A. W., Meaney, T. F.: *In Vivo* ^{31}P MRS study of human tumours in response to radiation therapy using a 1.5 T MRI System, in: NMR in Cancer (ed.) Allen, P., Pergamon Press (in press) 1986
90. Peterson, R. H. F., Meyers, M. B., Spengler, B. A., Biedler, J. L.: Cancer Res. *43,* 222 (1983)
91. Jesson, J. P., Meakin, P., Knissel, G.: J. Am. Chem. Soc. *95,* 618 (1973)
92. Ross, B., Marshall, V., Smith, M., Bartlett, S., Freeman, D.: Lancet, 641 (1984)
93. Clarke, G. D., Fatouros, P. P., Raynor, R., Montoui, J., Kishare, P. R. S.: J. Nuclear Medicine *26,* 828 (1985)
94. Bines, S. D., Tomasovic, S. P., Frazer, J. W., Boddie, A. W.: J. Surgical Res. *38,* 546 (1985)
95. Iwanik, M. J., Shaw, K. L., Ledwith, B. J., Yanovich, S., Shaw, J. M.: Cancer Res. *44,* 1206 (1984)

The Role of Immunoassay in the Clinical Laboratory

Eileen L. Nickoloff, Ph.D.

The Squibb Institute for Medical Research, Route 1 at College Farm Road, P.O. Box 191, New Brunswick, New Jersey 08903-9990

The technique of immunoassay has revolutionized clinical medicine over the past two decades. Compounds such as hormones, peptides and drugs, which could not be readily assayed, are now routinely studied.

With the advent of nonisotopic assays, which have in some cases replaced the earlier radio-immunoassays, confusion has arisen in the minds of many users as to the relative advantages and disadvantages of the different systems. Additionally, it is difficult to choose a method from the many which are available. Even after a method is selected, there are numerous problems which may arise, many of which are not readily apparent.

This article deals with the theory of immunoassay, method selection and quality control and is intended to give the reader a wider appreciation and understanding of this technique.

Progress in Clinical Biochemistry and Medicine, Vol. 3

1 Introduction[1-6]

The technique of immunoassay began over 20 years ago with the pioneering work of Berson and Yalow (Yalow subsequently won the Nobel Prize in 1977 for this effort). Today, immunoassays are used in clinical laboratories for the measurement of hundreds of drugs, hormones, steroids and other substances present in biological fluids in very low quantities. With the commercial availability of kits and instrument systems, clinical laboratories today have a rapid, convenient analytic method for measuring these substances. In many cases, immunoassay techniques replaced the earlier, much more cumbersome, chemical and chromatographic techniques.

The term "radioimmunoassay" was originally used by Berson and Yalow to describe what we know today as the competition of a radioactively labeled analog of the substance to be measured, and the substance itself (present in a biologic fluid), for a limited number of binding sites on a specific antibody. Today we have analogs which are labeled with non-radioactive compounds (thus these assays are termed "nonisotopic" immunoassays) as well as assays using, not a specific antibody, but an endogenous binding protein or receptor. Additionally a number of assays do not have the antibody in limited supply, and thus are not "competitive" immunoassays at all.

Often, these assays are referred to, not as immunoassays, but as ligand assays. Ligand comes from the latin "ligare", to bind, and thus refers to assay methods which involve the binding of a compound to a specific receptor. The prefix "radio" in a radioligand assay indicates that the label used on the ligand is a radioactive one and that this radioactive label will be used as the indicator for quantification of the assay. Obviously, however, radionuclides are not the only label that can be used in the ligand assay. If the ligand is attached to an enzyme, the assay is called an enzymoimmunoassay, or "EIA". The ligand may also be a fluorescent molecule, in which case the assay is called a fluoroimmunoassay, or "FIA". Other molecules with properties such as chemiluminescence have also been used in chemiluminoimmunoassay (CIA) systems.

It is important to understand however that these are all ligand assays in the sense previously described and that there will be a binding of the labeled ligand to a specific receptor or antibody. Whatever the nature of the label, the assays all share the same basic principles of specific, noncovalent binding of one reactant by another, and all involve immune principles. This is especially necessary to keep in mind in view of the conflicting advertising for the different isotopic and nonisotopic immunoassays which have proliferated today. While these assays may have different advantages and disadvantages, they are all based on antigen-antibody binding and therefore theoretically should all give the same results.

All competitive immunoassays are represented by the equation:

$$Ag + Ag^* + Ab \rightleftharpoons \underbrace{AgAb + Ag^*Ab}_{\text{"bound"}} + \underbrace{Ag + Ag^*}_{\text{"free"}}$$

In a competitive immunoassay, unlabeled antigen (Ag) from the patient's sera or urine and labeled antigen (Ag*) compete for binding to a limited number of antibody or other binder sites (Ab) to form a "bound" antigen-antibody complex. Unbound or "free" antigen will also be present due to the limited amount of antibody. As shown here, this is a heterogeneous reaction, in that the "bound" complex must be separated from the "free" antigen so that the different amounts of one or both fractions can be determined. While most radioimmunoassays are heterogeneous, requiring separation, nonisotopic assays may be homogeneous, where no separation of bound from free is required. This is because there are ways of changing the signal in a nonisotopic assay such that it is different when the labeled antigen is in the "free" or "bound" form. In an RIA, the gamma emission of ^{125}I (or any other radioactive label used) is the same whatever the environment of the ^{125}I.

There is another type of immunoassay, called a sandwich or labeled antibody reaction, which uses a labeled specific antibody rather than a labeled antigen. This will be further discussed later in this chapter.

Since the terminology of immunoassay can be confusing, an explanation of common terms used in immunoassay procedures follows.

1.1 Glossary of Abbreviations and Definitions

Accuracy. The degree to which the experimentally obtained value agrees with the true value.

Analyte. Substance to be measured by an assay system.

Antibody. An immunoglobulin produced in lymphoid cells in response to stimulation by a foreign substance (antigen or immunogen) and capable of reacting *in vitro* with this antibody.

Bo (Maximal Binding). An estimate of the maximal amount of labeled antigen that can react with the binding agent (e.g., antibody) under the conditions of the assay. Bo is determined in the absence of unlabeled antigen; therefore, it is equivalent to the binding of the zero standard.

Background Count. That portion of the counts recorded by the counter (or spectrometer) that is intrinsic to the measurement system when no labeled samples are being counted.

B/Bo. A ratio of the average net cpm of a sample (control, standard or unknown) to the average net cpm of the maximal binding (zero standard) tubes.

Binding. A chemical process of attraction and holding between molecules. In competitive radioimmunoassay, it is measured by the amount of radiolabeled or unlabeled antigen that reacts with antibody.

Binding Agent. The test reagent chosen to react specifically with the substance under test. In competitive radioimmunoassay the binding agent is usually an antibody or endogenous binding protein.

Bound. In a competitive radioimmunoassay, the fraction of radiolabeled antigen recovered or calculated to be in the ligand-binding agent (e.g., antigen-antibody) complex after separation of the bound and free fractions.

Chemiluminoimmunoassay (CIA). Is one in which the label on the antigen is a chemiluminescent molecule such as luminol.

Coefficient of Variation (% CV). A measure used when comparing data between different series of analyses. The comparison is made by dividing the standard deviation of the set by its mean and is usually expressed as a percent.

Competitive Binding Assay. An assay where the labeled and unlabeled ligand are simultaneously incubated with the antibody and compete for the antibody binding sites.

Competitive Protein Binding (CPB) Assay. A general term for any assay that employs a protein binder and where a labeled and unlabeled ligand compete for binding sites. Commonly, the protein binder is an endogenous binding protein.

Control Sample. Samples of known composition run to check the precision of the assay.

Counts per Minute (CPM). The average rate of nuclear disintegrations in a test sample perceived and recorded by a detector during a one minute interval. Used in RIA only.

Cross Reactivity. The degree to which a substance, usually chemically or physically similar to the substance being measured, reacts with the antibody in the assay system. Cross-reactivity is usually expressed as a percentage.

Curve Fitting. In competitive immunoassays, the plotting in various ways of the labeled antigen as a function of the concentration of analyte in the standards to determine which system gives the best representation.

Decay. The spontaneous disintegration of a radionuclide resulting in a decrease, with time, of the number of radioactive events in a sample.

Dilution. 1) The difference in concentration of test substances between points in a standard curve. 2) The addition of diluent to a test sample to a) match the concentration of the standards more closely or b) minimize the effect of interfering substances.

Dispenser or Dispenser/Diluter. A mechanical device to dispense reagents and/or dilute samples or standards.

Dose Response Curve. A plot of the dose of the analyte being tested versus the amount of labeled component, usually represented as a ratio (see Standard Curve).

Double Antibody Separation. A method of separating antibody – bound from unbound labeled antigen by using a second antibody which is directed against the first (or primary) antibody coupled to the antigen.

Dual Label. The use of two tracer species (usually different radioisotopes) in an assay system.

Efficiency. The percent of emissions a counter or spectrometer detects.

Enzymoimmunoassay (EIA). One in which the label on the antigen is an enzyme. The assay can be quantitated in an ultraviolet spectrophotometer. Enzyme immunoassays can be homogeneous or heterogeneous.

Fluoroimmunoassay (FIA). One in which the label on the antigen is a fluorescent molecule. The assay can be quantitated in a fluorometer.

Free. In a competitive immunoassay, the fraction of labeled antigen that is not bound to the antibody.

Gamma Emission. Energy originating from unstable atomic nuclei. Gamma rays are electromagnetic radiations of very short wavelength, which are the radioactive emissions most frequently used in radioimmunoassay.

Half-life. The time required for one-half of the radioactive nuclei present in a population to decay; cannot be changed by physical or chemical reactions.

Hapten. An incomplete antigen that cannot by itself induce antibody formation but can do so if coupled to a protein carrier and will react with the antibody so produced.

Heterogeneous Immunoassay. Is one in which the antibody bound and free fractions must be separated from one another before either one can be quantitated.

Homogeneous Immunoassay. Is one in which the antibody bound and free fractions do not need to be physically separated in order to be differentiated from one another.

Immunogen. A substance capable of inducing an immune response when introduced into a suitable animal host.

Immunometric Procedure. One in which an excess of labeled antibody is present. These assays are also known as labeled antibody or sandwich assays. This type of assay always results in a standard curve in which the number of counts in the precipitate is directly proportional to the amount of ligand present.

Interassay Variability. A measure of the reproducibility of the values of a sample between several separate runs of the assay.

Intraassay Variability. A measure of reproducibility of the values of a sample within a single run of the assay.

In-vitro Assay. Is a term often used to describe immunoassays. The phase in-vitro is used to differentiate it from in-vivo assays in which radioactivity is administered directly to the patient. In an in-vitro assay, radioactivity is mixed with biologic fluid from the patient, and is not administered to the patient.

Isotope. One of a group of nuclides of the same element having the same number of protons but a different number of neutrons than other isotopes of the same element.

KeV. Kilo electron volts = 1000 electron volts. A unit of energy applied to nuclear radiation particles and rays.

Kit. A complete set of calibrators and reagents pre-packaged for convenient performance of an assay. Includes a protocol for assay performance.

Label. A substance that is attached to or part of a ligand that permits its detectibility. In RIA, the label is a radionuclide; in EIA, an enzyme; in FIA, a fluorescent molecule.

Ligand. The substance that is bound; usually refers to the antigen.

Nonisotopic Immunoassay. Is one in which the label on the antigen is non-radioactive.

Nonspecific Binding (NSB). The binding of the antigen to assay reagents or the test tube in the absence of specific antibody.

Normal Range. The pair of numbers which are the 2.5 and 97.5 percentile points for the distribution of assay values for healthy persons. Can be further defined as REFERENCE RANGE or INTERVAL, which includes a statement of how the values were produced.

Percent Bound (% B). The percentage of labeled material bound to the antibody or receptor after separation compared to the total amount initially added to the tube.

Precision. The degree of agreement of repeated assays of a sample; usually expressed as a coefficient of variation.

Primary Antibody. The antibody that is directed towards the substance being assayed.

Protocol. A detailed description of the exact manner in which an assay is to be run.

Quality Control. In general terms the analytical and other steps that must be taken to insure that results of assays are reliable and representative for the true concentration of the material to be assayed.

Radioimmunoassay (RIA). A competitive binding assay where radiolabeled antigen and unlabeled antigen compete for specific binding sites on an antibody.

Receptor Assay. A type of competitive binding assay in which the primary binding agent is prepared from tissue and is not antibody. Receptors from the tissue specifically bind the ligand being assayed.

Run. A group of samples processed as a unit and comprised of standards, controls and test samples.

Sensitivity. The lowest concentration of test substance that can be differentiated from the zero standard.

Specificity. The degree to which interfering substances do not affect assay results. Usually applied to the antibody.

Standard. A set of samples containing known amounts of test substance, used to prepare a standard curve.

Standard Curve. A plot of activity measurement versus known concentrations of the test substance in the set of standards.

Therapeutic Range. Also referred to as optimal concentration range; that range of serum values for a drug within which most patients will exhibit a desired therapeutic effect.

Titer. Measurement of antibody concentration. For an immunoassay the titer is frequently defined in terms of the dilution of antisera that will bind 50% of added labeled ligand.

Total Counts (TC). The counts in a tube containing only that amount of labeled antigen added to all tubes in a run, but no other reagents.

Tracer. A test substance labeled with a marker, such as a radioactive isotope or a fluorescent compound, in such a way as to compete equally with unlabeled substance for antibody binding sites.

2 Components of the Immunoassay System

2.1 Binder[7–14)]

The binder in an immunoassay system may be an antibody, endogenous binding protein, or receptor. The binder is the component that is common to all immunoassays, and its properties to a large extent determine the quality of the assay. As will

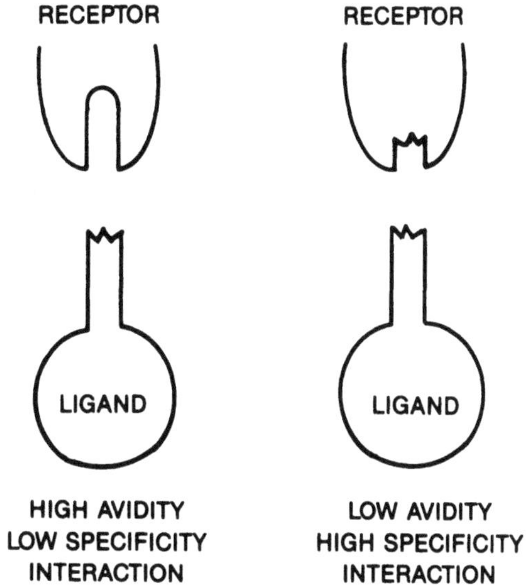

Fig. 1. Schematic representation of the concepts of avidity and specificity. A highly valid binding has large contact areas between receptor and ligand; however, there may not be high specificity (unique fit). There *is* unique fit in high-specificity binding, but it may not be strong

be discussed in the section on quality control, the quality of an assay depends on four factors: sensitivity, specificity, precision and accuracy.

Specificity refers to the ability of the binder to attract only the desired analyte and to be free of interferences from similar compounds. While specificity refers to the uniqueness of the binding, the term affinity is often used as a measure of the strength of the chemical attraction between the binder and the analyte. Figure 1 demonstrates the concepts of specificity and affinity.

Specificity (or its lack) may or may not be a problem, depending on the number (and concentration) of cross-reacting substances in the biologic fluid to be tested (this will be further discussed in Sect. 5.1). A lack of affinity, however, will always be a problem, since to measure the very low amount of analytes detected by an immunoassay, the binding energy between binder and analyte must be very high. Thus, the affinity of an antibody is often the limiting factor in immunoassay sensitivity.

2.1.1 Chemistry of Antibodies[15–36)]

Antibodies are a group of related proteins in blood, which are collectively called immunoglobulins (Ig). There are five main classes of immunoglobulin: IgG, IgM, IgD, IgA and IgE, all with a similar basic structure (Fig. 2).

Each class of immunoglobulin has separate activity: for example, the IgE class is responsible for allergic phenomena. The IgM class is the major immunoglobulin of several primitive species and generally occurs only in the earliest phase of immune response in higher animals. For immunoassay, the most important type of immunoglobulin is IgG, since this generally encompasses antibodies to bacterial, viral and other exogenous antigens.

All immunoglobulins have a pair of heavy chains, each associated with a light chain. The chains are arranged as shown in Fig. 2 for IgG, with disulfide bonds

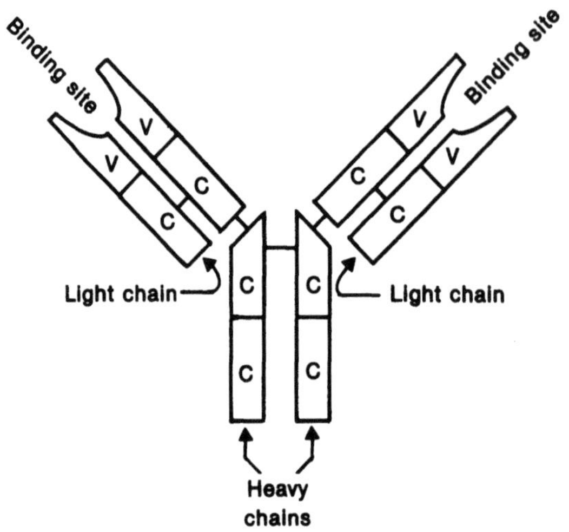

Fig. 2. Schematic representation of immunoglobulin G (IgG) structure. "C" are regions with constant structure, while "V" denotes the binding site regions, where variability makes it possible for the binding site to fit to its corresponding antigen

joining the chains, and a molecular weight of approximately 160,000 daltons. It is important to note that each IgG molecule can bind two antigen molecules, one on each of the binding sites located on the open ends of the Y-shaped molecule. Since these binding sites (called the variable portion of the chain) may differ in chemical composition, the specificity (and affinity) of the antibody is determined by the structure of this portion of the molecule.

Polyclonal antibodies for immunoassays are generally produced in animals such as goats, rabbits, sheep or guinea pigs. Since immunogenicity is generally linked to the molecular weight of the molecule, if the molecule has a molecular weight greater than 1000 to 3000 daltons, it can be injected directly into the animal. If it is below this weight it is considered a hapten, and must be linked to a protein such as bovine serum albumin or thyroglobulin before being injected into the animal. The production of immunoglobulin in response to stimulation with the immunogen generally follows a time course such that the antibody produced initially is of the IgM type. This response is very rapid and IgM antibodies may appear in plasma as early as 12 h after immunization. After about a week, however, the cells producing the IgM antibodies switch to synthesis of IgG antibodies. IgG production after a single stimulus usually reaches a peak after one to two months and then declines slowly. In general when monthly booster shots are given, these cause a rapid IgG response without any previous IgM induction.

Not every substance foreign to the animal's body will induce an immune response. Small molecules generally do not. Certain types of compounds are more immunogenic than others. Most large proteins and carbohydrates are immunogenic. However, the antibodies developed to the substance will bind only to a very limited portion of the antigen molecule. The antibodies produced in response to a immunogenic stimulus constitute only a small fraction of the total antibody population occuring in that animal. In most assays there is no need for separation of the particular fraction from the other antibodies which are present since the assay system only measures the binding of one particular antigen, and other antibodies present will not usually interfere in this binding. An antibody mixture of this type is known as a polyclonal antisera.

The antibody itself is contained in diluted animal sera. This degree of dilution in known as titer, which is a function of both concentration and affinity. Commonly it refers to the optimum antisera dilution for a particular assay. For routine assays this antisera dilution is selected on the basis of test sensitivity and range.

We have said that conventional antisera contain mixtures of antibodies. These mixtures vary from animal to animal. Each antibody is made by a different line of lymphocytes and their derived plasma cells. In 1975 a group lead by Cesar Milstein and Georges Köhler learned how to fuse mouse myeloma cells with lymphocytes from the spleen of mice immunized with a particular antigen. The resulting hybridoma cells expressed both the lymphocyte property of specific antibody production and the immortal character of the myeloma cells. Individual hybrid cells could be cloned and each clone produced large amounts of identical antibody to a single antigenic determinant. These highly specific antibodies, called monoclonal antibodies[37–46)], have proven to be a remarkably versatile tool in many areas of biological research in clinical medicine.

Preparations of monoclonal antibodies can be defined with respect to their characteristics, such as actual binding site and affinity for that site. Theoretically there are a number of advantages of monoclonal antibodies over polyclonal antibodies, especially in immunometric assays, where large volumes of antisera are required. It would be expected that specificity would be far improved over that of polyclonal antibodies, and this is indeed true. However, since the antibody is binding to only a small portion of the antigen, it is found that the sensitivity of pure monoclonal antibodies is generally less than is required for an immunoassay. For this reason mixtures of several monoclonal antibodies are often used in immunoassay procedures. Additionally, monoclonal antibodies are best prepared, like polyclonal antibodies, with a highly immunogenic antigen.

One of the real uses of monoclonal antibodies is in tumor marker assays, because of the advantage of monoclonal antibodies over polyclonal antibodies that a pure antigen is not required. Since the antibody clones will be screened and selected on the basis of desired characteristics, undesirable antisera strains can be eliminated at the screening step. Indeed, monoclonal antibodies are the only way to produce antibodies against tumor markers and other impure antigenic determinants.

2.2 Labeled Antigen

All immunoassays require a mechanism for determining the distribution between the antibody-bound and free fractions. For this reason, a known amount of highly purified labeled ligand is added to each tube in an immunoassay run. The label must be able to be measured accurately with great sensitivity.

2.2.1 Isotopic Labels[47–59)]

The most common isotopic label in use is iodine-125, which has a half-life of 60 days and a gamma emission at 35 keV. ^{125}I is readily counted in a gamma-counter. Other radioactive labels which may be used include tritium (^{3}H) and carbon-14, but these emit β-particles, and must be detected in a liquid scintillation counter, where the isotope is in direct contact with a scintillator solution (an aromatic compound which emits light when excited by ionizing radiation). Since common scintillators are usually only soluble in organic solvents such as toluene, this presents a variety of problems both in mechanics of counting and in radioactive waste disposal. For this reason, ^{125}I is the most popular radiolabel, and is used in a wide variety of applications.

In addition to providing the basis of quantification, the radiolabeled antigen, in an RIA, is used to provide a measure of the total ligand in the bound and free phases of the system. Therefore, it should react with the antibody in as similar a fashion as possible to the unlabeled antigen. Tritium is useful here because it is easily inserted into the antigen molecule in place of a normal hydrogen atom. Iodine-125 however, must usually be substituted onto another group such as tyrosine which is then connected to the antigen to form the radioactive label. Iodine may also substitute onto other amino acids, including histamine. While this may change the structure of the antigen, it still must behave similarly with the unlabeled antigen in the assay system. However, differences may arise because of the presence of the label on the molecule (including the group to which the label is attached). Additionally, differences may arise because of the introduction of impurities in the course of preparing the radiolabeled antigen.

It is also possible for the radiolabel to cause damage to the molecule to which it is attached. This is called radiolysis and may account for destruction of a part of the radiolabeled antigen with time.

A number of different iodination techniques are available including chloramine-T and lactoperoxidase. Different procedures must be used depending on the structure of the molecule to be iodinated as well as its stability.

2.2.2 Nonisotopic Labels

There are a variety of both real and perceived advantages to nonisotopic labels. These include:

- A prolonged shelf life, since the radiolabel is not subject to radioactive decay. Thus, problems of distribution and storage may be eliminated.
- There is no radiation hazard.
- It may be possible to have a homogeneous assay wherein the bound and free portions of the assay mixture do not require separation.
- There is no radioactive license required; nor is there any radioactive waste disposal.

However, there are also a number of disadvantages to nonisotopic labels. These include

- The assay may not be as sensitive as a radioimmunoassay and may not be able to detect analytes to as low a level in biological fluids.
- Substances with properties similar to the label may occur as normal components in a biologic fluid and may yield background noise which could severely limit both sensitivity and precision.

2.2.2.1 Enzyme Labels[60–83)]

Enzyme labels (used in an enzymoimmunoassay) are currently the most widely used of the nonisotopic labels. Because of the catalytic nature of enzyme activity, a single molecule of an enzyme may be responsible for the conversion of a number of molecules of substrate. The enzymes must be stable and have high activity at a pH which does not disturb the antigen-antibody reaction. Additionally, enzymes for homogenous systems must be very resistant to non-specific interference from serum. Enzymes which have been used as labels in heterogeneous EIAs include horseradish peroxidase, alkaline phosphatase, beta galactosidase and catalase. In homogeneous EIAs, lysozyme, malate dehydrogenase and glucose-6-phosphate-dehydrogenase are commonly used.

2.2.2.2 Fluorescent Labels[84–106)]

The potential sensitivity of fluorescence determination is very great although the high background noise which is seen often limits the sensitivity. The most commonly used fluorescent label is fluorescein isothiocyanate (FITC). Other molecules used are rhodamine and rare earth chelates.

2.2.2.3 Chemiluminescent Labels[107–121)]

Luminescence is associated with reactions which result in the emission of photons of light. The reaction may involve a synthetic chemical (chemiluminescence) such as the oxidation of luminol or acridinium compounds in the presence of H_2O_2 and a catalyst. End point detection is extremely sensitive and, like fluorescence, very rapid. However, technical difficulties have thus far prevented the widespread application of luminescent labels in binding assays. One problem is the quenching of 99% or more of the luminescence after conjugation to biologic materials.

2.3 Purified Ligand Standards[122–123)]

2.3.1 Purified Ligand

One of the requirements for the components of a binding assay is highly purified ligand to use as a standard. If highly purified material is freely available (such as it will be with some of the synthetic small peptide hormones and many of the steroid hormones), it can routinely be used for standardization, for preparation of tracer, and for immunization. In general, ligands which are prepared synthetically such as steroids, small peptide hormones and drugs present relatively few problems, while ligands which must be prepared from natural sources such as large protein hormones and oncofetal antigens may present considerable difficulty.

In many cases, synthetic material is not available and purified material extracted from human or animal tissues must be used. This is true in the case of the glycoproteins such as LH, FSH and TSH. While preparations of these are available from international agencies, their purity as judged by biological potency by weight unit is less than that of the most highly purified preparations reported by individual workers. In many cases, values obtained from preparations developed for use in bioassays differ significantly from results obtained by immunochemical means.

Ideally, the ligand used in an assay as both tracer and standard should be identical with the endogenous ligand which that assay is intended to measure. However, this is frequently not the case. There may be species differences in the use of non-human antigens in radioimmunoassays for human hormones. There may be tissue differences caused by the source of the material from the body; for example, growth hormone exists in a variety of different forms in the pituitary and circulating in blood. These forms have different biological activities and widely different potencies in immunoassays. There may also be multiple forms of the materials. Thus, insulin is synthesized as a larger precursor molecule called proinsulin. At the time of release this is broken down to yield the two peptide chain of insulin together with the connecting peptide chain (C-peptide). Under certain conditions proinsulin may be released into the circulation but it will not be measured by the majority of insulin assays. There may also be problems caused by metabolites of endogenous materials, which will be further discussed in the section on cross-reactivity. Because of differences between laboratories using different reagents and assay protocols, there will invariably be a wide range of different results for the same analyte between laboratories. There has thus been considerable interest in development of standards which could be used by all laboratories to normalize their assay results. The preparation and distribution of common standard materials of this type is performed by the World Health Organization as well as the National Pituitary Agency of the National Institute of Arthritis, Metabolism and Digestive Diseases (NIAMDD) in the United States and by several European agencies.

2.3.2 Standards

For a material to be considered as a standard, it should be available in large quantities, stable, pure, and it should, when possible, be a single molecular species. A primary standard obtained by one of the groups such as the World Health Organization can then be used as a source against which to calibrate different batches of standards used in a particular assay.

The purified ligand to be used as a standard in an immunoassay is usually dissolved in serum or a protein based pool such that it will compare with the matrix of the sample from the patient. Oftentimes problems with standards are due not to the analyte but to mismatching of the matrices between the patient sample and the matrix of the standards. It is difficult to develop a matrix for the standards that will match all variations in patient sera, where there may be differences in serum proteins, lipid concentrations, etc. Analyte-free serum is usually used as the matrix for standards. However, this serum has been treated with charcoal or ion exchange resin to remove the analyte, and this treatment may affect the properties of the matrix compared to the patient sera. There may also be differences seen if the standards are made up in sera and the patient sample used in plasma or urine.

2.4 Separation Techniques

In a heterogeneous immunoassay, the antibody-antigen reaction must always be followed by a step which separates the antibody bound from the free antigen, since ordinarily only one of these will be counted. The efficiency of separation is never perfect. One measure of separation efficiency or degree of overlap is given by the non-specific binding (NSB) tube. This measures how much labeled antigen is bound to material other than specific antibody, and is performed by running a tube in the assay with all assay components except antibody. Any binding of material is then due to non-specific causes.

There are a large number of separation techniques[124–134], most of which vary with time, temperature, protein concentration and pH of the solution. Additionally, with some types of separation, trapping or non-specific sequestration of bound material in the free phase (or vice versa) may occur.

Some separation techniques are designed to bind or adsorb free antigen while others bind or adsorb the antigen-antibody complex. Historically, charcoal was one of the first separation systems used. Charcoal functions by adsorbing free antigen, which co-precipitates with the charcoal to the bottom of the tube. Charcoal, because it has a large surface active area, is non-selective and may also adsorb some of the bound antigen-antibody portion. Therefore, it is often treated with dextran to block some of the active sites. This material, dextran-coated-charcoal (DCC) is more selective for the binding of the smaller free antigen molecules.

Silica and talc have also been used, and function by adsorbing the free phase. Since they tend to bind more weakly then charcoal, stripping is less pronounced.

A third method of separation involves ion exchange resins to attract the free antigen. The resins can be in the form of strips, columns or suspensions. These eliminate the need for centrifugation and thus simplify the technique. However, there are limitations. The ion exchange method is generally sensitive to variations in pH as well as protein and salt concentrations. The method also displays moderate to great time dependance. However, separation efficiency can be quite good.

There are several methods for separation which depend on molecular size. One of these is molecular sieve or gel filtration chromatography, where the reaction mixture is poured thru a gel column which selectively excludes large molecules. Another method utilizing molecular size is called salting out. Certain salts such as ammonium sulfate, or organic materials such as polyethylene glycol, will tie up water molecules in solution when added in high concentrations. This causes the less soluble larger proteins and antibody-bound antigens to precipitate out of the mixture. While inexpensive, fast and simple, these methods tend to be rather non-specific.

An old but reliable technique is the double antibody or second antibody method of separation. This technique uses a second antibody produced from a different animal species than the first to precipitate the complex. The second antibody specifically attaches to the primary antibody, which is attached to the antigen. This results in large polymer-like aggregates of molecules which can be easily centrifuged. The reaction is highly specific and gives good separation efficiency. While it is not protein dependant, the pellet formed is relatively frail.

Another type of separation is known as solid phase. These include methods such as antibody coated to the inside of test tube walls, or coated to chips, beads or other solid supports. In these methods, there is little or no centrifugation and thus fewer pipetting steps. One problem with these methods is that non-specific binding (NSB) can not be determined because it is not possible to have an antibody-free tube as a reference.

A variation on solid phase technology is called the sandwich solid phase system. It is generally applied to relatively large antigens which have multiple binding sites. This technique uses labeled antibody instead of labeled antigen. A solid support such as a bead is prepared with an excess of specific antibody attached. Next, sample is added and reacts with the support-bound antibody. Finally, labeled antibody is added which may be the same or different than the antibody added first. The labeled antibody attaches to other binding sites on the antigen, creating an antibody-antigen-antibody sandwich. The beads are then washed and counted. This type of method is often known as an immunometric method and can be one-site or two-site depending on whether the antibodies added are the same or different.

Additional methods of separation include double antibody solid phase (DASP) where double antibody and solid phase methods have been combined into a single technique, and magnetizable particle separations, where the antibody is attached to a magnetic particle and, after the reaction has taken place, the particle is precipitated out of solution by applying a magnetic field.

3 Types of Immunoassay Systems

The main differences between different types of immunoassay systems is the nature of the labeled ligand, and whether or not the system is heterogeneous or homogeneous. Following is a brief discussion of several different types of immunoassay systems, including some positive and negative aspects of each.

3.1 Radioimmunoassay[135–143)]

In radioimmunoassay the label on the antigen is radioactive (usually I-125, but occasionally tritium or carbon 14). The advantages and disadvantages of radioimmunoassay are both derived from the radioactive nature of its label. Radioactivity can be detected with great sensitivity, and radioactive labels provide a clear distinct signal for detection. However, the signal is the same whether the antigen is attached to the antibody or is free. As a result, radioimmunoassays require separation of bound from free antigen, and thus have been difficult and expensive to automate. Additionally there are emotive problems involved with the use of radioactive isotopes and real problems in the distribution and disposal of radioactivity.

Radioimmunoassay is currently the most sensitive immunoassay method. However, there are some cases where the exquisite sensitivity obtainable in RIA is not required. While many interferences have been reported for RIAs, most of these will affect all immunoassays, whatever the label. There is, however, one specific interference seen in RIAs, and that is endogenous contamination of the patient sample with radioactivity from another source, usually imaging studies performed in Nuclear Medicine departments.

In general, the method and extent of the interference of various radioisotopes in RIA is largely determined by the separation procedure of the particular assay, since the contamination will be distributed between antibody "bound" and "free" fractions. However, the problem is more complex than first realized because the degree of contamination of the patient sample is also dependent on a number of other factors including:

1. which radioisotope the patient received, and how much of it was given
2. the rate of clearance of this radioisotope from the patient's serum
3. when the blood sample was drawn in relation to when the radioisotope was administered
4. when the laboratory runs the blood sample (since further radioactive decay will occur *in vitro*) and
5. sample size used in the assay

Since the radioactivity in the patient sample is subject to physical decay, it is possible that an erroneous result will be reported which, upon reassay, cannot be verified. The value obtained may be falsely elevated, falsely lowered or unaffected. This problem was first reported in 1977, but is often unrecognized in Clinical Chemistry laboratories unfamiliar with the use of *in-vivo* radioisotopes. It is recommended that nonspecific binding tubes be included for all patient samples or that samples be screened for endogenous radioactivity before assay.

The types of errors that have been reported are most prevalent with technetium 99 m (^{99m}Tc), the most commonly used radioisotope in Nuclear Medicine studies, with a half-life of 6 h and an energy of 140 KeV. However, interference has also been reported with gallium 67 (^{67}Ga), which has a physical half-life of 78 h and a gamma emission at 93 KeV, as well as thallium 201 (^{201}Tl), with a half-life of 73 h and a Mercury X-ray emission at 68–80 KeV. In many cases, due to Compton scatter, these radioactive emissions cannot be separated from the ^{125}I peak.

3.2 *Enzymoimmunoassay*[144–150)]

In enzymoimmunoassay the label contains an enzyme rather than a radioactive isotope. EIA was developed to avoid the problems of radioactive waste disposal and limited shelf life of radiolabeled reagents. While radioactive labels are useful because they have an inherently favorable signal to noise ratio, enzymes are attractive alternatives because each enzyme molecule generates many molecules of product when exposed to substrate, thereby amplifying the signal and producing a favorable signal to noise ratio. Sensitivity in a properly constructed EIA can be comparable to an RIA; however, in practice it is usually less sensitive.

There are several types of enzymoimmunoassay. All have in common two key features: the recognition of the analyte under investigation is achieved with the use of highly specific antisera, and the detection of this analyte is made by use of an enzyme labeled conjugate. Enzyme immunoassays can be either homogenous or heterogenous. Most homogenous assays utilize the EMIT® technique (Syva, Palo Alto, CA). Heterogenous EIAs are usually enzyme linked immunosorbent assays (ELISA).

Homogenous assays are distinguished from other EIAs by their ability to measure the product of the reaction in-situ, thereby eliminating the requirement for separation of bound from free label. Reaction of the enzyme-linked conjugate with the antibody either induces a conformational change in the tertiary structure of the enzyme or results in a blocking of the active site. Whatever the mechanism, the end result is that the enzyme is either fully active or is partially inhibited, in proportion to the concentration of analyte in the reaction mixture.

In contrast, heterogenous EIAs use an enzyme-labeled conjugate in which the catalytic activity behaves identically irrespective of whether or not the antigen is bound to antibody. ELISA assays can be configured to measure antigen or antibody concentrations.

In general, there are more interferences reported with homogenous EIAs than with heterogenous ones. In any EIA, the most common specific interference is that caused by the interaction of the enzyme label and the serum matrix of the patient sample. Serum is comprised of a mixture of proteins, carbohydrates, lipids, and other substances which may act as potential effectors of enzyme activity. The separation step in a heterogenous EIA reduces interactions with endogenous activators or inhibitors.

3.3 Fluoroimmunoassay[151–156)]

Like EIAs, FIAs may be either homogeneous or heterogeneous. The heterogeneous assays are quite similar to RIAs, with the exception that the label is usually tagged with a fluorescein derivative. The sensitivity of an FIA can, like other immunoassay techniques, be related to the signal to noise ratio (S/N), where the noise can be instrumental or chemical. Thus, it is critical that factors affecting noise, both in the instrument and the assay reagents, be minimized. However, the sensitivity of an FIA can be drastically affected by interferences from the matrix used for the standards and samples, since there may be endogenous fluorophores in the sample matrix which either fluoresce at the same wavelength as the label or quench fluorescence in the label. This problem can usually be resolved by either utilizing a heterogeneous assay format or by diluting the sample so that the interfering substance is also diluted.

Fluorescein-labeled haptens may also bind to serum albumin, a binding which can result in a quenching of the fluorescence and an increase in fluorescence polarization. This non-specific binding effect may be seen in the presence of normal human serum; it is, however, generally not seen with fluorescein-labeled protein antigens and/or antibodies.

The problem of endogenous fluorophors is most often seen in icteric, hemolyzed or lipemic sera. In serum, bilirubin is highly bound to albumin and in disease states, where serum bilirubin concentrations are elevated, it is not possible to discriminate fluorescein fluorescence from that of bilirubin. The hemoglobin in a lysed patient sera may cause a similar effect. In lipemic sera, high triglyceride levels may lead to a light-scattering signal requiring correction before assay. All of these problems are generally resolved by either a separation step or sample dilution. This would also be the case for FIAs performed on urine samples, where there may be high levels of interfering compounds.

3.4 Other Types of Immunoassay Systems[157–165)]

There are a number of other labels which may be used as substitutes for radioisotopes in immunoassay systems. These include chemiluminescent precursors, bacteriophages, fluorescence quenchers, coenzymes, inhibitors, various particulate metal atoms and stable free radicals. New methods are described virtually every month. These will not be further discussed in this monograph; however, excellent review articles are available.

4 Immunoassay Method Selection[166–168)]

With all of the methods available today, how does a laboratory go about selecting the best method for their particular situation? It is necessary to consider the resources available to the laboratory. Can all forms of immunoassay be performed or is only a gamma counter or UV spectrophotometer available? What is the concentration of the analyte to be measured? Is automated equipment for a particular type of immunoassay available? The answers to these questions will help the established laboratory determine which type of immunoassay is best for a particular application. However, for a laboratory just being set up, the choice must be based on which future assays will be desired and which ones are available for each of the different types of methods.

Once the type of immunoassay has been determined, it must then be decided which of the available methods within that class would be best. It is important to differentiate between a kit and components, both of which may be commercially available. A kit includes all the major reagents required for the assay as well as a specific protocol to follow (either manual or automated). Components are individual reagents available from a number of manufacturers and may or may not include a specific analytical procedure. Components may be mixed (i.e., antibody from one supplier and labeled antigen from another) or supplemented by reagents prepared in-house. Additionally, all reagents required for an assay may be prepared in-house. However, most clinical laboratories tend to use either complete kits or components due to the lack of time and/or expertise required to prepare individual reagents. If a laboratory has a particular automated instrument which can only use reagents from a sole supplier, no choice need be made. However, those who can choose must first decide what the most important parameters in kit selection are for themselves.

Most laboratories feel that precision and accuracy of the method are of prime importance; however, other factors also play a major role. These include:

- *Performance Time and Ease.* A busy clinical laboratory where technologists are often performing many different tasks may have difficulty in using a method which demands split-second timing for incubations or intricate technical operations such as extraction of the sample. While there are cases where there is no simple alternative, very often one can be found which is acceptable.
- *Sample Size.* The serum sample size required for most manual immunoassays ranges from 10 to 200 μL. A laboratory performing a large volume of pediatric studies may wish to select a method with a sample volume on the low side of this range (especially if the sample will be run in duplicate).
- *Special Equipment Needed.* Some manual assays demand equipment which may not be readily available, such as a refrigerated centrifuge, special-volume pipets, water baths, etc. It is important to understand the importance of particular items and to decide if they can be easily borrowed or must be purchased. If this is not possible, use of a particular kit may be precluded.
- *Service from the Company.* In purchasing a finished kit (or even components), a laboratory will be paying significantly more than if it prepared these reagents in-house. Part of this price is due to the cost of development and part to the cost of quality control. But for this higher price, a laboratory can and should expect a response when one has a problem. Most companies are truly concerned about customer problems and will work with the customer to resolve them.
- *Cost.* In these days of cost-consciousness, this is obviously an important factor. But the cost of most kits depends to a great extent upon the amount of reagents purchased each month and does not usually significantly differ from company to company (although it may differ drastically between different types of immunoassays, and for manual *vs* automated methods). A laboratory must weigh the quality of the final result and the simplicity of the method, as well as the factors previously mentioned, against a difference in price.

How can these factors be simply and easily evaluated? One of the quickest ways is to carefully study the package inserts from several different companies. These will indicate accuracy and precision, and will aid in assessing such parameters as equipment which will be needed and whether the procedure appears applicable to an individual laboratory's needs. Other parameters can also be considered at this time. For example, some of the factors that a laboratory may wish to consider in evaluating performance time and ease might include the separation method of the particular test, the sample volume required, incubation times and temperatures, reagent reconstitution required, and the stability of these reagents once reconstituted. Additional factors to be looked at may include the washing procedure, calculations, the number of standards provided, whether or not internal controls are provided, etc.

Reagent reconstitution may be a very major point here. Some kits require reconstitution of all reagents; others, only some or none. Once reconstituted, reagents may have differing stabilities, for the life of the kit or only a week or other limited period of time. To a laboratory performing a volume of assays which will use up an entire kit in one run, stability of a reagent once reconstituted makes no difference. However, if a laboratory purchases a kit of 100 tubes and plans to do four runs of

25 tubes each over a four-week period, the stability of the reconstituted reagent, if it is only one week, will be a major factor. For a laboratory performing an assay only infrequently, a non-isotopic method with a long shelf-life may be preferred.

After reviewing the package inserts and narrowing the choice down several kits which appear to have good possibilities, it is time to ask for an evaluation kit and actually perform the test. This initial evaluation should include an assessment of precision, accuracy (when possible), parallelism, and agreement of the method with the laboratories current method (or some alternate mechanism for a preliminary indication of clinical correlation).

4.1 Precision

Precision may be classed as either intraassay (within assay) or interassay (between assays). Precision of a method is usually calculated using the percent coefficient of variation (% C.V.) which is one standard deviation divided by the mean, or

$$\frac{1\ \text{S.D.}}{\text{mean}} \times 100 = \%\,\text{C.V.}$$

Evaluation of intraassay precision is usually performed by running 15 or more duplicates of the same sample as unknowns within a single run. The concentration of each sample is calculated and the mean, standard deviation and percent coefficient of variation are determined. For interassay precision, several runs are performed, using aliquots of identical samples (usually in high, mid and low ranges). The values obtained from the different runs are then calculated to give the mean, standard deviation and percent coefficient of variation.

The interassay coefficient of variation is a measure of repeatability from run to run. If a specific value on a sample is obtained today, the same or a similar value should be obtained with the same sample next week or next month. Interassay variation is almost always greater than intraassay variation and indicates the reproducibility of the assay.

There are no firmly stated rules for acceptable precision of an assay, since to an extent the obtainable precision will depend upon the concentration of the analyte being measured in the biologic fluid. For T4 assays, for example, in which the analyte concentration is measured in μg/dl, less than a 5% coefficient of variation for intraassay variability and less than 10% for interassay variation over a one-month period would be expected. For triiodothyronine (T_3) assays, because the analyte concentration is measured in ng/dl (at a concentration approximately 70 times less than T4 in sera), about 10% for intraassay variation and 10 to 15% for interassay variability might be reasonable as a first estimate.

Obviously, different estimates of precision may be obtained at the high, mid and low ranges of a particular assay. A TSH assay in which the normal range is 0 to 7 μIU/ml may give repetitive values for a low control of 1.0, 1.5, and 0.5 μIU/ml. While this is acceptable variability clinically, the percent coefficient of variation will be large and is meaningless in this case.

Table 1. Intraassay precision of eight kits for T_3 RIA, using a normal control sera. Suggested normal ranges are from package inserts

Kit	Suggested Normal Range (ng/dl)	Mean + Standard Deviation (ng/dl)	% Coefficient of Variation (% CV)
A	64–215	115 ± 10.1	8.8
B	100–230	120 ± 6.8	5.7
C	70–200	137 ± 13.0	9.8
D	60–160	91 ± 2.3	2.5
E	106–188	109 ± 7.4	6.8
F	65–200	58 ± 5.2	9.0
G	59–200	100 ± 20.0	20.0
H	80–220	133 ± 15.0	11.3

In evaluating the precision of different methods, a laboratory must look at both the mean obtained for the sample being tested and the % C.V. Obviously, if the normal ranges of the kits being studied are different, this must also be taken into account.

Table 1 shows results obtained with eight different T3 RIA kits, running 18–30 duplicates of a commercially prepared normal-range control serum within a single run of each kit. It can be seen that the % C.V. varies from 2.5 to 20.0%, reflecting differences in precision among the kits. More significantly, however, is the fact that the means are changing to reflect different assay ranges, with Kit F giving the lowest value and Kit C the highest. These changes cannot be completely explained by the differences in suggested ranges, and indicate that more than precision must be considered.

4.2 Accuracy

The reason that different means are obtained is that each of these kits shows different accuracy (or recovery of analyte). It might be expected that the same specimen, run by different methods, should give the same value (or at least that differences would be reflected in differences in the normal ranges). This is often not the case and a recovery study can help to clarify the situation.

Recovery studies are performed by preparing solutions with known concentrations of the analyte (usually high, mid, and low levels) and assaying these solutions by the methods under evaluation. Obviously, the analyte must be available in a pure form that can be added to the solution quantitatively. Thus, it is possible to perform recovery studies on most drugs and on other molecules such as T3 and T4; it is impossible with analytes such as PTH or CEA where a pure sample cannot be obtained.

The matrix used to dissolve the analyte for a recovery study is also of critical importance. In general, it should be compatible with the matrix used in the kit standards, assuming this is protein-based. Human serum may be used (for compounds not endogenously present, like drugs) or stripped serum for compounds which naturally occur in serum. Especially in the latter case, the system must be checked for matrix effects (the standards acting differently from the sample); a blank solution consisting of the matrix without added analyte should also be used to check for compatibility.

There are a number of ways to calculate recovery. It is usually expressed in terms of either a) the total analyte expected and the total found or b) the total amount of analyte added to the matrix and the amount recovered. While both methods give approximately the same values, b) is more sensitive to imprecision at extremities of the standard curve. Method a) will usually give slightly higher values and is not as sensitive to non-specific binding effects.

Thus:

$$\text{a)}\ \frac{\text{total analyte found}}{\text{total analyte expected}} \times 100 = \%\ \text{recovery}$$

$$\text{b)}\ \frac{\text{total analyte recovered}}{\text{total analyte added}} \times 100 = \%\ \text{recovery}$$

Analyte added = calculated analyte in aliquot of spiked sample added to matrix

Analyte recovered = total found – concentration of analyte in blank

If the concentration of analyte in the blank is 0, similar values would be obtained for methods a) and b). For example, if a sample for recovery study for a T3 assay is prepared to contain 60 ng/dl of T3, but when running the assay a value of 30 ng/dl with a 0 blank is found, the recovery would be 30/60 × 100% or 50%, and it would be expected that the assay would give falsely low values in this range. On the other hand, if the assay of the same 60 ng/dl sample gave 90 ng/dl with a 0 blank using a different kit, the recovery would now be 90/60 × 100% or 150%, and it would be expected that falsely elevated values would be obtained in this range using this method.

When the eight kits evaluated for precision (Table 1) were now analyzed for accuracy, using prepared solutions containing 60, 150 and 300 ng/dl T3, the results

Table 2. Intraassay accuracy of eight kits for T_3 RIA

Kit	% Recovery 60 ng/dl	% Recovery 150 ng/dl	% Recovery 300 ng/dl	Normal Control Serum, Mean (ng/dl)	Elevated Control Serum, Mean (ng/dl)
A	107	117	127	115	245
B	98	107	108	120	235
C	114	118	118	136	264
D	84	89	99	91	235
E	120	109	121	109	278
F	40	60	50	58	–
G	101	128	150	100	440
H[a]	25	56	75	133	225

[a] Probably invalid, due to high blanks

shown in Table 2 were obtained. These data demonstrate that the differences in the mean obtained with the control sera in the precision test correlate in most cases with the recovery of the assay in that range. For example, Kit F, which gave the lowest value for the normal control serum, also had the lowest percent recovery of any kit tested. Similarly, Kit C, which gave the highest value for the normal control serum, had the highest percent recovery. In the case of Kit H, the BSA added to the matrix used in the recovery study may have interfered in the assay, as evidenced by very high blanks. A solution without BSA should be prepared to test this kit.

It can be seen that no single kit gives ideal results. In a recovery study values of from 90 to 110 percent in each assay range would be expected. Kit B, therefore, might be a good selection of a method for this assay. Kit C, which has higher recoveries that are consistent in every range, would also be acceptable, but it would be expected that the normal range for Kit C would be higher than that of the other kits. The worst situation would be a kit such as G, for which the recoveries at different analyte concentrations are different. This kit gave a value for the normal control serum which agreed reasonably well with that of the other kits because the recovery in that range was somewhere between 101 and 128%. However, this kit had the highest percent recovery (150%) in the high range, and it can be seen that the value for the elevated control sera obtained with this kit is much higher than that obtained by any other method.

4.3 Parallelism

Another important parameter to evaluate in immunoassay is parallelism. Parallelism testing demonstrates that the dilution curve of an elevated sample will parallel the standard curve for the assay by a constant factor and that there is linearity over the range of the standard curve. The purpose of a parallelism study in immunoassay procedures is two-fold:

1. If parallelism is seen in the assay, a correct answer can probably be obtained from a diluted sera. In other words, if a sample has a concentration above the highest standard on the standard curve, it can be diluted to give an answer which can be read from the standard curve.
2. Non-parallelism suggests that the sample and the standard are not reacting with the antibody in a similar fashion; one of the reasons for this may be that an interfering substance is present in the sample. In addition, non-parallelism may occur because the labeled ligand has been damaged so that it no longer reacts with the antibody in the same manner.

It is important to realize that demonstrating parallelism does not guarantee that no interfering substance is present. However, with a commercial kit, it is a good test to help assure that there are no serious problems. Parallelism studies are usually performed by preparing several dilutions of an elevated or suspect sample and running these as unknowns in the assay. It is important to use a diluent which will not cause any matrix effects and, for this reason, the same considerations which apply to the choice of a diluent for a recovery study must also be evaluated in this case.

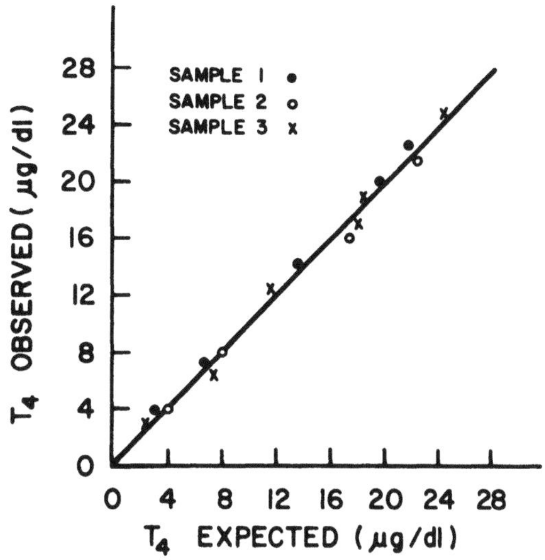

Fig. 3. Parallelism calculated from plot of expected versus observed values. Sample 1 (●) has a slope of 1.06, a y-intercept of 0.22, and a correlation coefficient of 0.999; Sample 2 (○), a slope of 0.93, a y-intercept of 0.40, and a correlation coefficient of 0.997; and Sample 3 (×), a slope of 0.97, a y-intercept of 0.12, and a correlation coefficient of 0.993

There are a number of simple methods for plotting parallelism, two of which will be outlined here.

Method. 1. The expected value (x axis) may be plotted against the observed value on the y axis (Fig. 3). The observed value is read from the standard curve, and the expected value is calculated as follows:

$$\frac{\text{undiluted serum value}}{\text{dilution factor}}$$

Note that the expected value for the undiluted serum is taken to be the same as the observed value from the graph. If, for example, a parallelism plot was made for T4 and the undiluted serum was found to be 24 µg/dl, the expected value for the 1:2 dilution would be calculated to be 12 µg/dl; the 1:4 dilution 6 µg/dl, etc. Theoretically, this curve should be a straight line running through the origin, with a slope of 1.0. Deviation from this straight line can be mathematically calculated.

Method 2. Parallelism can also be depicted by *arbitrarily* assigning a value to the undiluted serum sample such that this point will not be superimposable upon the standard curve. The concentrations of the diluted samples are then derived from this arbitrary concentration. For example, if the undiluted sample is arbitrarily assigned a value of 24 µg/ml, the 1:2 dilution would be 12 µg/ml, the 1:4 dilution would be 6 µg/ml, etc. These assigned values are then plotted on the same graph paper as the standard curve (Fig. 4) against the B/Bo (or B/T value) previously determined for each dilution. If the two curves are parallel, the assay is considered valid.

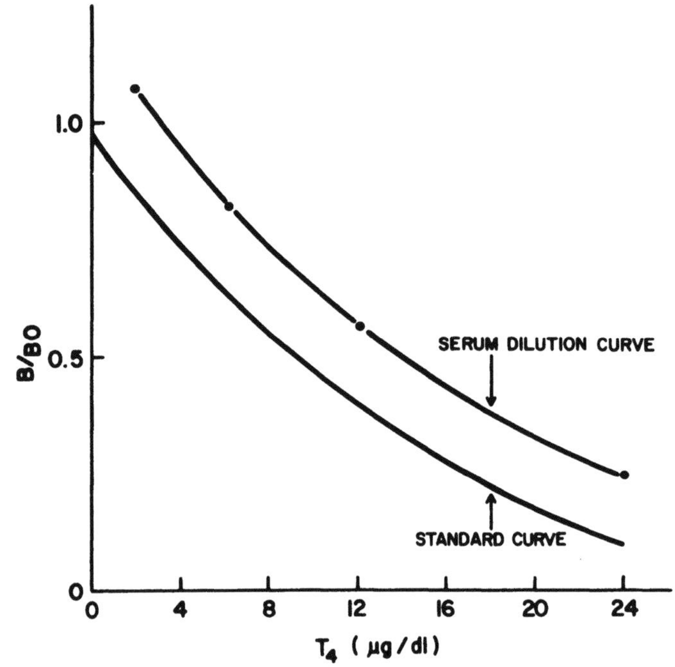

Fig. 4. Parallelism determined by plotting an arbitrarily assigned value for an undiluted sample against the standard curve

4.4 Differences Between Recovery and Parallelism

While it may seem that parallelism and recovery studies are very similar (and indeed they are), there are some basic differences between them. While recovery studies primarily measure the accuracy of the assay system, parallelism tests the validity of the method, and demonstrates that the sample and standard ligands will similarly displace radioactive ligand from antibody binding sites. In recovery studies, the amount of analyte present in the sample studied is determined by independent means (gravimetrically, spectrophotometrically, etc.). In parallelism, the undiluted sample is arbitrarily or experimentally assigned a value. It is thus possible for an assay to show good parallelism, but unacceptable recovery.

4.5 Final Criteria

By using the criteria outlined in the previous sections, the decision-making process in selecting a kit is simplified. After a review of the package insert to see how a procedure will fit into a particular laboratory's routine, several kits can be selected for further evaluation. Studies of precision, accuracy, and parallelism should then enable the laboratory to make a knowledgeable choice of a single kit.

5 Problems in Immunoassays[169–223)]

5.1 Cross-Reactivity

While immunoassay is a very sensitive method for assaying many substances, it is not without problems. A major one is in knowing exactly *what* is being measured in the assay, since similar compounds and/or metabolites may also react with the assay antibody. Generally, these interfering substances are structurally similar to the antigen of interest, and this interference is called *cross-reactivity*.

The cross-reacting compound may be a precursor of the compound we wish to measure (proinsulin, for example, may interfere in an immunoassay for insulin). It may also be a metabolite of the analyte, as is the case with dihydrodigoxin, a digoxin metabolite. Drugs with similar structures may cross-react; digitoxin, which is a drug almost identical to digoxin in structure, usually cross-reacts to varying extents in digoxin assays. Sometimes the situation is more complex; many package inserts for digoxin assays show a 0% cross-reaction for spironolactone. These studies are performed by taking serum from a normal subject and adding a known quantity of spironolactone. However, if serum is taken from a patient who is actually being treated with spironolactone, there is an interference in the digoxin assay. This is because canrenone, a spironolactone metabolite, *does* cross-react, even though the parent spironolactone doesn't.

While % cross-reactivity may be calculated in several ways, most package inserts list the cross-reactivity measured at 50% inhibition. In this method, several concentrations of cross-reactant are run in the assay as unknowns. A dilution curve is prepared from these different concentrations, and plotted on the same graph paper as the analyte standard curve. The concentration at which the cross-reactant curve reaches a B/Bo of 0.50 (or 50% inhibition) is noted, and the % cross-reactivity is:

$$\frac{\text{Wt. of analyte required to displace 50\% of the } ^{125}\text{I radiolabel}}{\text{Wt. of interfering substance required to displace 50\% of the } ^{125}\text{I label}}$$

This can be better understood in the following example for estriol (Fig. 5) where the estriol standard curve is plotted on the same graph as different dilutions of epiestriol, a cross-reacting compound. From this graph, the percent cross-reactivity would be:

$$\frac{22 \text{ ng/ml (wt. of estriol at B/Bo of 0.5)}}{240 \text{ ng/ml (wt. of epiestriol at B/Bo of 0.5)}}$$

or 4.8%.

It is critical when using this method to convert the units of cross-reactant to the same units as the concentration of the analyte. Thus both estriol and epiestriol are shown in ng/ml.

There are several problems in using this method. One is that in some cases of low cross-reactivity, the cross-reactant curve may be almost horizontal and may never

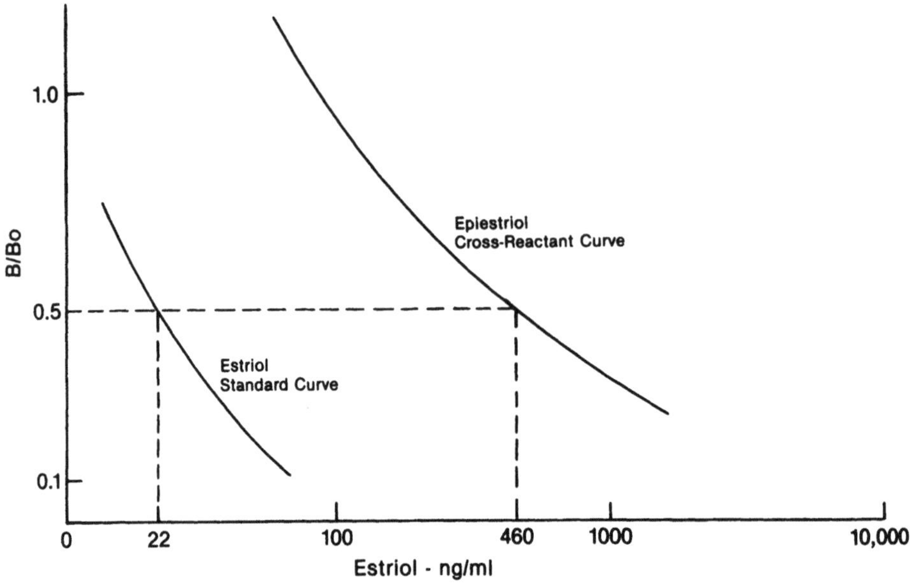

Fig. 5. Cross-reactivity curve of epiestriol against the estriol standard curve. From this curve, the cross-reactivity of epiestriol at 50% inhibition would be (22 ng/ml)/(460 ng/ml) × 100%, or 4.8%

reach a B/Bo as low as 0.5. In these cases, a solution of the interfering substance is analyzed and the concentration value (as analyte) is divided by the actual concentration of the interfering substance in the solution and multiplied by 100. For example, T2 at a concentration of 50 g/dl (50,000 ng/ml) yielded an assay value equivalent to 350 ng/dl of triiodothyronine. Therefore, the percent cross-reactivity is:

$$\frac{350 \text{ ng/dl}}{50{,}000 \text{ ng/dl}} \times 100 = 0.7$$

Another problem that may arise is that the cross-reaction dilution curve may not parallel the analyte standard curve, and the percent cross-reactivity will be different at the 50% inhibition level than at other (and possibly more physiologically meaningful) levels. For this reason, 50% inhibition cross-reactivity values must be interpreted with care.

It must also be remembered that a high cross-reactivity is not necessarily bad. For example, much has been made over the cross-reactivity of dihydrodigoxin in digoxin kits. A study of serum dihydrodigoxin levels after digoxin administration found dihydrodigoxin in sera of only 3 of 100 patients studied, at levels from 0.18 to 0.36 ng/ml. Since therapeutic levels of digoxin are 0.5–2.0 ng/ml, even a high cross-reactivity of dihydrodigoxin would not significantly change the digoxin assay results.

Cross-reactivity, while a very important parameter, must be understood and carefully considered in order to know whether high values are indeed significant, and for which interfering compounds low values are essential.

5.2 Other Interferences

A major problem encountered in most laboratories today is that there is often little control over the sample which is received. Most laboratories see only a tube of blood and it is sometimes easy to forget that it came from a patient. It must be remembered that problems in drawing the specimen, or problems inherent in the specimen itself, may drastically alter the results obtained with that sample.

There is much emphasis by manufacturers today on stressing the simplicity of kit methods for immunoassays. However, immunoassay procedures involve complex, kinetic reactions and things can go wrong in an assay that are not readily apparent. This is particularly true in considering the patient sample to be used.

Some parameters involving the sample that may affect assay results are:

- the sample may be hemolyzed
- the quantity of sample may be insufficient
- the sample may contain anticoagulant

5.2.1 Hemolysis

A hemolyzed sample may affect the assay because many antigens bind to red cells. In fact, the first assay for T3 Uptake used red cells as the binding agent rather than an antibody or resin binder. The effect of hemolysis usually cannot be corrected by measuring nonspecific binding. The best course of action is to establish the degree of interference by hemolysis on a particular assay. This can be done by drawing two tubes of blood from a willing subject, hemolyzing one and running the other normally. The effect of hemolysis on the assay can then be observed and, if it is significant, additional samples can be taken from different subjects so that a valid conclusion can be drawn.

Obviously, if hemolysis does affect assay results, the laboratory must attempt to obtain a fresh, unhemolyzed sample. However, it is often true that a subsequent sample cannot be obtained either because the patient has been discharged or because medication has been started or stopped on the patient such that a current sample is no longer valid for the assay. The laboratory must then try to determine the extent to which this interference may affect the assay.

If in evaluating the effect of hemolysis in a T4 assay it is found that an unhemolyzed sample gives a value of 11.9 μg/dl and a hemolyzed sample taken from the same patient at the same time gives a value of 9.6 μg/dl, this difference should then be confirmed by running additional samples. Subsequently, when a hemolyzed sample is received in the laboratory, the result may be reported. However, on the report sheet, the laboratory would note that the sample was hemolyzed and that hemolysis may lower the result on the assay by up to 20%. This is not ideal, but at least it gives the physician an indication that interference was present and some idea of the scope of that interference.

5.2.2 Insufficient Sample Size

A second possibility is that an insufficient volume of sample may be received by the laboratory. If the assay calls for 500 μl of sample and only 100 μl is received, the question arises how to best handle this. In most immunoassays, the worst thing to do is to simply use a smaller sample size than called for in the protocol, because most of these tests are very protein dependent. By using a smaller sample size, less protein is added. If the analyte being tested is endogenous in serum, the sample should be brought up to the required volume by using the "0" standard from the kit as the diluent. If sufficient "0" standard is not available, a treated serum free of the constituent being tested should be added. In the case of some analytes such as T_3 and T_4, serum can be purchased which has been stripped of these components. In the case of drug assays, sera from a normal individual who is not taking the drug may be used as the diluent.

5.2.3 Anticoagulants

Oftentimes, although the protocol for an assay calls for serum, plasma will be sent to the laboratory instead. Serum is usually obtained in red top tubes, free of anticoagulants, whereas plasma will contain various anticoagulants. The best way to handle this type of problem is to take two tubes of blood from a willing subject, one being serum and one containing the anticoagulant in question. The two tubes can be run simultaneously in the assay and the results compared. If the anticoagulant does not appear to have an effect on the result, the result obtained from the patient sample may be reported. However, if the results are different, one cannot calculate the effect of the anticoagulant (since it might be different at different analyte concentrations, etc.) and one should report that an improper sample was received and not report a result on this sample.

5.2.4 Stability

Another variable in the sample is the stability of the analyte in the serum. Most laboratories have had a physician call and explain that an extremely important sample was taken from a patient, and the patient was released from the hospital, but unfortunately, the physician forgot to send the sample to the laboratory and has "carried it in his pocket for a week". Whether or not this sample is still valid depends upon the stability of the particular constituent being tested in serum. Many thyroid hormones are very stable in serum and can survive even the drastic treatment described. However, a sample to be assayed for folate could not be used after such treatment. The stability of each analyte in serum must be tested.

Certainly an indication of stability can be obtained by reviewing the literature. However, it is a simple matter to take two tubes of blood from a willing subject (who by now is probably getting very tired of having you take blood from him or her!), running one sample immediately and allowing the second tube to stand at room temperature for up to one week. At the end of this time, the sample can be spun down, the assay run, and the results compared.

6 Quality Control[224–235]

The purpose of quality control in the clinical laboratory is to insure that reliable results are obtained 100% of the time. In the clinical laboratory mistakes must be avoided insofar as is possible since patients may be treated on the basis of results obtained. A commonly used method for evaluating quality control is the Levey Jennings Chart.

6.1 Levey Jennings Chart

This chart was first described for clinical laboratory use by Levey and Jennings in 1950 and has gained popularity in most laboratories as a simple way to graphically visualize control serum data. While the design is familiar to most laboratorians some do not realize that it is simply an extension of a Gaussian normal distribution curve laid over on its side, as shown in Fig. 6.

While daily results can be expressed numerically, the visual impact of a Levey Jennings Chart carries far more weight. The chart represents the area of the Gaussian curve which falls between the ±2 standard deviation limits. These may be classified as allowable limits of variation or confidence limits, although they are not truly parameters of in- or out-of-control situations.

In using the Levey Jennings chart, the two limit lines drawn at the ±2 standard deviation levels are not to be construed strictly as demarcation lines between "in" and "out of control" situations. These lines are arbitrarily drawn to delineate the area under a Gaussian curve in which 95.5% of daily recovered values will fall if the test procedure is operating correctly and if random scatter is observed. Approximately once every 20 tests the test values will fall outside those two limits. Since this is the number of values that will fall in the third standard deviation area by chance

Fig. 6. The Levey-Jennings chart as an extension of a Gaussian normal distribution curve laid over on its side

alone, it cannot be said to indicate a definite out-of-control situation just because it appears on that portion of the chart. Values plotted outside of the two limits must be interpreted in context with the preceding values obtained.

In simple terms, the ±2 standard deviation values suggest the range between which any value is said to be equivalent to the mean value, insofar as the test procedure is able to reproduce the mean value. The occasional point falling outside the two standard deviation limits may or may not be significant, depending on the context. If the outliers occur more frequently then once in 20 successive assays or if there are two successive points outside the same limiting line, then investigation should be carried out to determine the cause of the bias.

It must also be understood that even though all values may fall inside the ±2 standard deviation limits, the assay may still not be in control. If six or more successive points fall above or below the mean value, the system is biased and hence out-of-control. The figure of six is empirical but has been shown to be practical for the busy laboratory. Mere chance alone will not allow this occurance, and something has occurred to change the values obtained.

In addition to Levey Jennings charts, many laboratories plot parameters such as total counts (especially in an radioimmunoassay), non-specific binding, and concentration of analyte at various points of binding. These actually include maximum binding, or Bo, and the concentration at 50% Bo/T, 20% B/T and 80% Bo/T. While these points are easy to evaluate for assays in which the standard curve is manually plotted or is printed by the computer, some nonisotopic immunoassays do not print standard curves and thus these important quality control parameters can not be determined.

6.2 Maximum Binding

Perhaps the most important tube in the assay in terms of quality control is the maximum binding tube (Bo). Counts from this tube are usually divided by total counts to give Bo/T. Ideally, the Bo/T will be about 50%.

Maximum binding serves as an indication of:
- Antibody Affinity
- Antibody Capacity
- Labeled Antigen Integrity
- Changes in Assay Conditions
- Kit Storage Stability
- Gross Undercentrifugation (in a heterogenous assay)

As a general rule, Bo/T values will decrease as a given set of reagents age; however, this is not always the case, especially in non-isotopic immunoassays. The limit to the shelf life of a set of reagents is generally the immunoreactivity life of the labeled antigen. With time, the antigen may degrade due to radiolysis (in an RIA), chemical alterations and structural relationship degradations. The primary antibody will no longer recognize the antigen in a denatured state so it is not bound as readily or at all. Hence maximum binding decreases and since the total counts are independent of biochemical action, Bo/T also decreases.

6.3 Concentration at 50% Bo/T

The concentration at 50% Bo/T is important to monitor since this point is usually on the most sensitive and most linear part of an immunoassay curve. Since this is the most stable part of the curve, values should change little from assay to assay. When kits are designed, the 50% Bo/T point is usually established to be in the most diagnostically important part of the curve.

6.4 Concentrations at 20% and 80% Bo/T

The concentrations at 20% Bo/T and 80% Bo/T generally encompass the usable limits of precision on a curve. Outside these points, sensitivity and thus precision is generally decreased. Within them is the most sensitive, linear and precise portion of the curve.

6.5 Normal Range[236–246)]

The most important consideration of any immunoassay is how well it correlates with the clinical condition of the patient. One parameter which is often neglected is determination of normal range.

There is a great deal of confusion prevalent about the concept of normal range at the current time. In statistics, the normal range describes the usual or prevailing condition; in medicine, it has come to refer to a healthy person, with departure from the normal being viewed as ill health. In many cases, normal ranges are extracted from literature (or package inserts) in which little information is given on the population studied or the particular definition of "normal" used in the selection of this population. A number of attempts have been made to replace the term "normal range" with the more relevant "reference values" or "reference intervals", which define the parameters under which subjects were chosen for the study. When a laboratory conducts its own studies, reference data are usually generated by analyzing results from studies on readily available "healthy" subjects. These results are then used to derive reference intervals which most laboratories call the "normal range". These values may not reflect assay differences caused by illness.

It is important that laboratories do establish their own normal range for two reasons. The first is that each laboratory has a bias on every test it performs. This bias is due to the different pipets or automated equipment used in the laboratory, the temperature at which the assay is conducted, and other variables which may affect results. Additionally, the normal range is best determined from the same population from which patients will be taken. Although it is not as true now as it used to be, there can be differences in some assay values caused by diet and other factors. For example, in the days before the use of iodized salt, thyroid function values were very different in populations at the seashore and those inland.

Today, the normal range provided by a laboratory is geared to 95% limits. This means that approximately 1 in 20 so-called "normal" individuals will have a value outside this range. Most physicians will consider a result normal if it is inside these limits, suspicious if it is somewhat outside these limits, and abnormal if it is considerably outside.

Most laboratories, if they establish a normal range at all, will select so-called normal employees, medical students, physicians, or whomever can be coerced into donating a sample. Sera from these "normal volunteers" will then be run and the normal range or reference interval is all too often established by assuming a normal distribution of the values obtained and then using ± 2 standard deviation limits to specify the upper and lower reaches of the normal range. There are several things wrong with this method. Selection of subjects for a normal range is of critical importance because this range will be affected by how the sample of healthy persons is chosen for the study. If is thus important to realize, for each particular assay, what factors may influence the normal range. With some assays, for example, the sex of the subject will cause differences in the normal range. Age may be important, as may race, weight or diet of the subject. With some assays the time of day the specimen is obtained may make a difference. An example of this is found with cortisol, which has a strong diurnal variation and gives higher values in the morning than at night. There may also be day-to-day variations, such as may be caused by the menstrual cycle, which will cause ferritin assays to be different in pre-menopausal women at different times of the month.

6.5.1 Statistical Methods

Obviously, the validity of the statistical method will depend upon the number of subjects chosen for study. Initially, at least 120 normal samples are required for statistical validation. Very few laboratories are able to assay sera from this many so-called normal subjects. In establishing the normal range, it is imperative to duplicate as closely as possible the actual conditions that will prevail once the test is introduced to the laboratory, and thus build interassay variability into the normal range. Therefore, sera from normal subjects should be assayed in several different runs over a two or three week period of time using different technologists and different lots of reagents, if possible.

It is also very important to simulate routine conditions using routine data reduction techniques. Thus, if it is planned to use a computer for data reduction after the test is introduced into the laboratory, the same computer program should be used to generate assay values used in the establishment of the normal range.

Once the sera have been collected and assay values obtained, it is necessary to analyze the data. There are complicated computer methods for doing this. We will briefly discuss three simple methods which may be less valid than some of the computer methods but which are generally acceptable for routine clinical use. The first step should be to plot the data and determine whether or not the distribtuion is indeed Gaussian or non-Gaussian. If the distribution is Gaussian, then the mean $\pm$ two standard deviations can be used as the normal range. However, many statisticians today believe that most distributions of normal individuals are non-Gaussian.

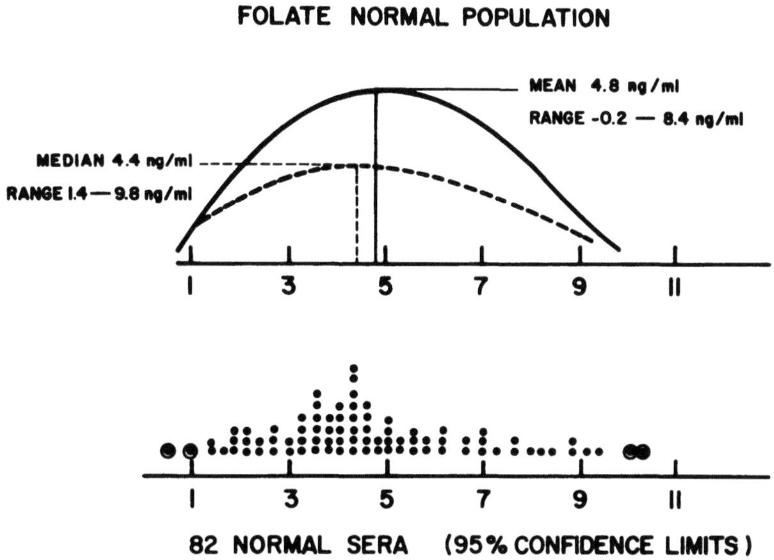

Fig. 7. Plot of folate values obtained from sera of 82 normal patients. Because distribution is non-Gaussian and normal range as determined from mean ± 2 standard deviations is invalid, the histogram method should be used to determine the normal range used in this example

Other methods must be used if the distribution is non-Gaussian. The easiest of these is the histogram method. Here, data are arranged in the histogram form and the lowest 2.5% of the values from the lower end of the distribution and the highest 2.5% of the values from the upper end of the distribution are eliminated. The range of the values remaining is the normal range. This method can only be used if the distribution is continuous, that is, if there are no gaps in the distribution (Fig. 7). If the data used in generating Fig. 7 were analyzed using the Gaussian distribution (mean ± 2 S.D.), a normal range of − 0.2 to 8.4 ng/ml would have been obtained. By using the histogram method, a normal range of 1.4 to 9.8 ng/ml is obtained. The histogram can be plotted as a simple dot graph as in Fig. 7, or as a bar graph (Fig. 8).

Another way of determining the normal range is by using the percentile estimate method, which is carried out by ranking the assay results to be analyzed in their order of magnitude. The sample 2.5 percentile is the rth ordered sample from the low end value, where $r = 0.025\ (n + 1)$, and n = the number of observed values; that is, the number of samples. The corresponding estimate of the 97.5 percentile is the rth ordered sample value from the high end. For most values of n, r is not a whole number, and one must interpolate between the two ordered sample values whose ranks are nearest and on each side of r. As an example of this method, assume that serum amylase levels were performed on 130 normal males aged 20 to 29. For this group, the lowest 10 values ranked in order were 47, 55, 57, 61, 61, 61, 62, 64, 65 and 66. The highest ten ranked in order were 114, 114, 114, 118, 122, 126, 128, 130, 131 and 153. The percentile estimate r would equal 0.025 (130 + 1) or 3.28. The 2.5 percentile is thus the 3.28th value from the low end, or 58.12, which is rounded off to 58, since results are reported only in whole units. Similarly, the 97.5

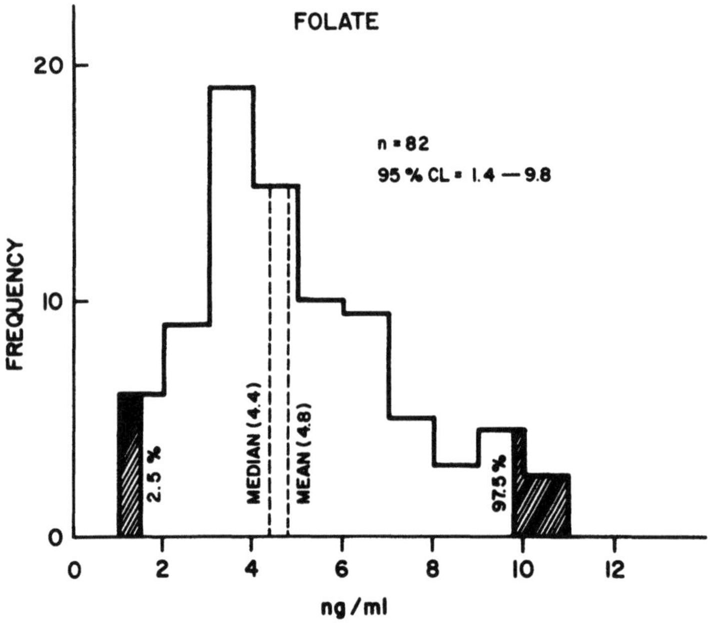

Fig. 8. Histogram bar-graph plot of data from Fig. 7 shows concentration vs frequency

percentile is the 3.28th value from the high end, i.e., 129.44, rounded off to 129. The normal range for this sampling would thus be 58 to 129. The reliability of the percentile estimate method improves with increased sample size. With 100 samples, the difference in precision between the result obtained using the Gaussian standard deviation and the percentile method is relatively small, if the sample has a Gaussian distribution. An advantage of the percentile estimate method is that it can be used for any distribution of patient values, even those non-Gaussian distributions for which the histogram method is not applicable.

6.5.2 Reference Ranges and Intervals

While substitution of the term reference range for that of normal value appears to be trivial, on closer examination this change can be seen as an important step toward establishing a scientific basis for clinical interpretation of the laboratory data.

DYBKAER[245)] has used the term reference value with the following meaning: "A set of values of a certain type of quantity obtainable from a single individual or a group of individuals corresponding to a stated description." This description must be spelled out and available if others are to use the reference values. Thus, one must specify age, sex, race, time of day, previous diet and exercise, posture, etc., of the subjects used for this study.

It should be emphasized that the term reference value denotes the entire set of test results that is obtained from a reference population. When statistical calculations are used to establish percentile limits, the result is designated a reference interval.

7 Conclusions

It has been the intent of this monograph to define the advantages and disadvantages of immunoassay for the clinical laboratory and to further define advantages and disadvantages of the different types of immunoassays for the particular laboratory situations. It must be emphasized that no assay which does not give clinically correlatable results is of use in the clinical laboratory, no matter how precise those results may appear.

Immunoassays are an extremely valuable resource but they become even more valuable once their limitations are understood and can be dealt with.

8 References

General References: 1–6

1. Chard, T.: An Introduction to Radioimmunoassay and Related Techniques, New York, Elsevier Biomedical Press 1982
2. Thorell, J., Larson, S.: Radfoimmunoassay and Related Techniques, St. Louis, CV Mosby 1978
3. Moss, A., Dalrymple, G., Boyd, C.: Practical Radioimmunoassay, St. Louis, CV Mosby 1976
4. Freeman, L., Blaufox, D.: Radioimmunoassay, New York, Grune & Stratton 1975
5. Geitner, A., Bering, N.: RIA in the Clinical Laboratory, American Society for Medical Technology 1977
6. Maggio, E. T.: Enzyme Immunoassay, Boca Raton, CRC Press 1980

Antibody Production: 7–14

7. Hartigan, P. J.: Irish Vet. J. *35,* 143 (1981)
8. Dekruyff, R. H., Siskind, G. W.: Cell Immuno. *49,* 90 (1980)
9. Moldow, R. L., Fischman, A. J.: Peptides *3,* 989 (1982)
10. Hindawi, R. K., Read, G. F., Riad-Fahmy, D.: J. Immunol. Methods *37,* 57 (1980)
11. Shahani, S. K., Bordekar, A. D., Shah, G. V., Sheth, A. R., Raoi, S. S.: J. Immunol. Methods *23,* 91 (1978)
12. Krause, V.: Scand. J. Clin. Lab. Invest. *39,* 533 (1979)
13. Pratt, J. J., Woldring, M. G., Boonman, R., Kittikool, J.: Eur. J. Nucl. Med. *4,* 159 (1979)
14. Hosoda, H., Miyairi, S., Kobayashi, N., Nambara, T.: Chem. Pharm. Bull. *28,* 3369 (1980)

Antigen-Antibody Binding: 15–36

15. Sidki, A. M., Landon, J., Rowell, F.: Clin. Chem. *30,* 1348 (1984)
16. Hosoda, H., Kobayashi, N., Nambara, T.: Chem. Pharm. Bull. *31,* 953 (1983)
17. Nordblom, G. D., Webb, R., Counsell, R. E., England, B. G.: Steroids *38,* 161 (1981)
18. Hosoda, H., Kawamura, N., Nambara, T.: Chem. Pharm. Bull. *29,* 1969 (1981)
19. Nordblom, G. D., Counsell, R. E., England, B. G.: Lig. Quart. *2,* 34 (1979)

20. Allen, R. M., Redshaw, M. R.: Steroids *32*, 467 (1978)
21. Mori, M., Wilber, J. F.: J. Lab. Clin. Med. *103*, 223 (1984)
22. Cook, C. E., Seltzman, T. P., Tallent, C. R., Wooten, J. D.: J. Pharmacol. Exp. Ther. *220*, 568 (1982)
23. Rodwell, J. D., Tang, L. H., Schumaker, V. N.: Mol. Immunol. *17*, 1591 (1980)
24. Budd, R. D., Leung, W. J., Yang, F. C.: Clin. Tox. *17*, 383 (1980)
25. Pratt, J. J., Woldring, M. G., Boonman, R., Kittikool, J.: Eur. J. Nucl. Med. *4*, 159 (1979)
26. Creighton, T. E.: Biochemistry *19*, 4308 (1980)
27. Muller, R.: J. Immunol. Methods *34*, 345 (1980)
28. Rorstad, O. P.: J. Immunoassay *4*, 49 (1983)
29. Mason, P. A., Law, B., Moffat, A. C.: J. Immunoassay *4*, 83 (1983)
30. Butler, J. E.: Antibody-antigen and antibody hapten reactions, in: Enzyme Immunoassay (ed.) Maggio, E. T., p. 5, Boca Raton, CRC Press 1980
31. DeGomez, M. S., Frei, A., Vitins, P., Cekan, S. Z.: Ann. Clin. Biochem. *18*, 281 (1981)
32. Geiseler, D., Ritter, M.: Anal. Chem. *54*, 2062 (1982)
33. Vallejo, R. P., Bogus, E. R., Mumma, R. O.: J. Agric. Food Chem. *30*, 572 (1982)
34. Budd, R. D.: Clin. Tox. *18*, 773 (1981)
35. Loveridge, N., Bitensky, L., Chayen, J.: J. Immunoassay *2*, 89 (1981)
36. Budd, R. D., Yang, F. C., Utley, K. O.: Clin. Tox. *18*, 317 (1981)

Monoclonal Antibodies: 37–46

37. Polin, R. A., Wasserman, R. L.: Laboratory Management, October, 33 (1983)
38. Gordon, D. S.: MLO, July, 53 (1983)
39. Sevier, E. D.: Am. J. Med. Technol. *48*, 651 (1982)
40. Secher, D. S.: Immunol. Today, July, 22 (1980)
41. Yelton, D. E., Scharff, M. D.: Am. Sci. *68*, 510 (1980)
42. Hunter, W. M.: Monoclonal antibodies in clinical chemistry, in: Proceedings of the Royal Society of Edinburgh, p. 293, 1982
43. Kung, P. C., Chang, T.-W., Zurawski, V. R.: Monoclonal antibodies in clinical investigations, in: Clinical Biochemistry, p. 89, New York, Academic Press 1984
44. Sevier, E. D., David, G. S., Martinis, J., Desmond, W. J., Bartholomew, R. M., Wang, R.: Clin. Chem. *27*, 1797 (1981)
45. Milstein, C.: Eur. J. Biochem. *118*, 429 (1981)
46. Gordon, D. S.: MLO, July, 53 (1983)

Radioiodination: 47–59

47. Corrie, J. E. T., Ratcliffe, W. A., Macpherson, J. S.: J. Immunol. Methods *51*, 159 (1982)
48. Sturgeon, C. M., Beynon, L., Seth, J., Chisholm, G. D.: Ann. Clin. Biochem. *20*, 112 (1983)
49. Bolton, A. E.: Irish Vet. J. *35*, 230 (1981)
50. Hamilton, R. G., Button, T. M.: J. Immunoassay *1*, 435 (1980)
51. Hnatowich, D. J., Layne, W. W., Childs, R. L., Lanteigne, D., Davis, M. A., Griffin, T. W., Doherty, P. W.: Science *220*, 613 (1983)
52. Abdulwahid, A. M., Yousif, S.: Intl. J. Appl. Radiat. Isot. *31*, 450 (1980)
53. Stanczyk, F. Z., Goebelsmann, U.: J. Steroid Biochem. *14*, 53 (1981)
54. Chisolm, G. M., Sila, C. A., Hmiel, S. P.: Anal. Biochem. *111*, 212 (1981)
55. Seevers, R. H., Counsell, R. E.: Chem. Rev. *82*, 575 (1982)
56. Hughes, J. P., Tanaka, T., Gout, P. W., Beer, C. T., Noble, R. L., Friesen, H. G.: Endocrinology *111*, 827 (1982)
57. Evans, E. A.: J. Radioanal. Chem. *64*, 9 (1981)
58. Corrie, J. E. T., Ratcliffe, W. A., MacPherson, J. S.: J. Immunol. Methods *51*, 159 (1982)
59. Youngblood, W. W., Kizer, J. S.: Strategies for the preparation of haptens for conjugation and substrates for iodination for use in radioimmunoassay of small oligopeptides, in: Methods in Enzymology, p. 435, New York, Academic Press 1983

Enzyme Labeling: 60–83

60. Kennedy, J. H., Kricka, L. J., Wilding, P.: Clin. Chim. Acta *70*, 1 (1976)
61. Peters, K., Richards, F. M.: Ann. Rev. Biochem. *46*, 523 (1977)
62. Levy, H. R., Inguli, J., Afolayan, A.: J. Biol. Chem. *252*, 3745 (1977)
63. Rowley, G. L., Rubenstein, K. E., Huisjen, J., Ullman, E. F.: J. Biol. Chem. *250*, 3759 (1975)
64. Kurzer, F., Douraghi-Zadeh, K.: Chem. Rev. *67*, 107 (1967)
65. Nakane, P. K., Pierce, G. B., Jr.: J. Histochem. Cytochem. *14*, 929 (1967)
66. Erlanger, B. F.: Pharmacol. Rev. *25*, 271 (1973)
67. Van Weemen, B. K., Schuurs, A. H. W. M.: FEBS Lett. *24*, 77 (1972)
68. Van Weemen, B. K., Schuurs, A. H. W. M.: Immunochemistry *12*, 667 (1975)
69. Exley, D.: The Influence of heterology and homology on enzyme immunoassay, in: Enzyme Labelled Immunoassays of Hormones and Drugs (ed.) Pal, S. B., Berlin, Walter de Gruyter 1978
70. Rubenstein, K. E., Schneider, R. S., Ullman, E. F.: Biochem. Biophys. Res. Commun. *47*, 846 (1972)
71. Comoglio, S., Celeda, F.: J. Immunol. Methods *10*, 161 (1976)
72. Rajkowski, K. M., Cittanova, N., Desfosses, B., Jayle, M. F.: Steroids *29*, 701 (1977)
73. Rowley, G. L., Rubenstein, K. E., Weber, S. P., Ullman, E. F.: Abstracts, 172nd Amer. Chem. Soc. Meeting, Washington, D.C., American Chemical Society (1976)
74. Exley, D., Abuknesha, R.: FEBS Lett. *79*, 301 (1977)
75. Ullman, E. F., Yoshida, R., Blakemore, J., Maggio, E. T., Leute, R. K.: Biochim. Biophys. Acta *567*, 66 (1979)
76. Kitagawa, T., Aikawa, T.: J. Biochem. *79*, 233 (1976)
77. Keller, O., Rudinger, J.: Helv. Chim. Acta *58*, 531 (1975)
78. Avrameas, S.: Immunochemistry *6*, 43 (1969)
79. Mannik, M., Downey, W.: J. Immunol. Methods *3*, 233 (1973)
80. Van Weemen, B. K., Schuurs, A. H. W. M.: FEBS Lett. *15*, 232 (1971)
81. Van Weemen, B. K., Schuurs, A. H. W. M.: FEBS Lett. *43*, 215 (1974)
82. Nakane, P. K., Kawaoi, A.: J. Histochem. Cytochem. *22*, 1084 (1974)
83. Nakane, P. K.: Ann. N.Y. Acad. Sci. *254*, 203 (1975)

Fluorescent Labeling: 84–106

84. Soini, E., Hemmila, I.: Clin. Chem. *25*, 353 (1979)
85. Neurath, A. R., Strick, N.: J. Virol. Methods *3*, 155 (1981)
86. Coons, A. H., Creech, H. J., Jones, R. N.: Proc. Soc. Exp. Biol. Med. *47*, 200 (1941)
87. Nakamura, R. M.: Lab. Res. Methods Biol. Med. *3*, 211 (1979)
88. Smith, D. S., Al-Hakiem, M. H. H., Landon, J.: Ann. Clin. Biochem. *18*, 253 (1981)
89. Quattrone, A. J., O'Donnell, C. M., McBride, J.: J. Anal. Toxicol. *5*, 245 (1981)
90. Jolley, M. E., Stroupe, S. J., Schwenzer, K. S.: Clin. Chem. *27*, 1575 (1981)
91. Ullman, E., Khanna, P.: Methods Enzymol. *74*, 28 (1981)
92. Kam, J., Yoshida, R.: U.S. Patent No. 4,252,783 (1979)
93. The, T. H., Feltkamp, T. R.: Immunology *18*, 865 (1970)
94. Blakeslee, D., Baines, M.: J. Immunol. Methods *13*, 305 (1976)
95. Hassan, M., Landon, J., Smith, D. S.: FEBS Lett. *103*, 339 (1979)
96. Chadwick, C. S., McEntegard, M. G., Nairn, R. C.: Immunology *1*, 315 (1958)
97. Miller, J. N., Lim, C. S., Bridges, J. N.: Analyst. *105*, 91 (1980)
98. Bailey, M. P., Rocks, B. F., Riley, C.: Ann. Clin. Biochem. *21*, 59 (1984)
99. Handley, G., Miller, J. N., Bridges, J. W.: Proc. Anal. Dir. Chem. Soc. *16*, 26 (1979)
100. Kronick, M. N., Grossman, P. D.: Clin. Chem. *29*, 1582 (1983)
101. Hendrix, J. L.: Clin. Chem. *29*, 1003 (1983)
102. Wilmott, N. J., Miller, J. N., Tyson, J. F.: Analyst *109*, 343 (1984)
103. Thakrar, H., Miller, J. N.: Anal. Proc. *19*, 329 (1982)
104. Burd, J. F., Carrilo, R. J., Fetter, M. C.: Anal. Biochem. *77*, 56 (1977)
105. Lovejoy, L., Holowka, D. A., Cathou, R. A.: Biochemistry *16*, 3668 (1977)
106. Soini, E., Kojola, H.: Clin. Chem. *29*, 65 (1983)

Chemiluminescence Labeling: 107–121

107. Woodhead, J. S., Weeks, I.: Pure and Appl. Chem. *57,* 523 (1985)
108. Whitehead, T. P., Kricka, L. J., Carter, T. J. N.: Clin. Chem. *25,* 1531 (1979)
109. Williams, D. C., Huff, G. G., Seitz, W. R.: Anal. Chem. *48,* 1003 (1976)
110. Schroeder, H. R., Yaeger, F. M.: Anal. Chem. *50,* 1114 (1978)
111. Kohen, F., Pazzagli, J. B., Kim, J. B.: FEBS Lett. *104,* 201 (1979)
112. Schroeder, H. R.: Luminescent immunoassays and binding assays monitored by chemiluminogenic labels, in: Luminescent Assays: Perspectives in Endocrinology and Clinical Chemistry, Serio, M., Pazzagli, M. (eds.), p. 129, New York, Raven Press 1982
113. Messeri, G., Martinazzo, G., Tommasi, R. in: Luminescent Assays Perspectives in Endocrinology and Clinical Chemistry, Serio, M., Pazzagli, M. (eds.), p. 207, New York, Raven Press 1982
114. Boguslaski, R. C., Schroeder, H. R., in: Bioluminescence and Chemiluminescence, DeLuca, M., McElroy, W. D. (eds.), p. 283, New York, Academic Press 1981
115. Campbell, A. K., Davies, C. J., Hart, R.: J. Physiol. *306,* 3P (1980)
116. Simpson, J. S. A., Campbell, A. K., Woodhead, J. S., in: Bioluminescence and Chemiluminescence: Basic Chemistry and Analytical Applications, DeLuca, M., McElroy, W. D. (eds.), p. 673, New York, Academic Press 1981
117. Woodhead, J. S., Simpson, J. S. A., Weeks, I., in: Monoclonal Antibodies and Development in Immunoassay, Albertini, A., Ekins, R. (eds.), p. 135, New York, Elsevier-North Holland 1981
118. Woodhead, J. S., Weeks, I., Campbell, A. K., in: Luminescent Assays: Perspectives in Endocrinology and Clinical Chemistry, Serio, M., Pazzagli, M. (eds.), p. 147, New York, Raven Press 1982
119. Schroeder, H. R., Osborn, D. D., Boguslaski, R. C., in: Immunoassays: Clinical Laboratory Techniques for the 1980s, p. 189, New York, Alan R. Liss 1980
120. Schroeder, H. R., Boguslaski, R. C., Carrico, R. J., in: Methods in Enzymology 57, p. 424, New York, Academic Press 1978
121. Simpson, J. S., Campbell, A. K., Ryall, M. E. T.: Nature *279,* 646 (1979)

Serum Standards: 122–123

122. Jawad, M. J., Wilson, E. A.: Clin. Chem. *27,* 1993 (1981)
123. Stuart, M. C., Ellis, S., Gowlland, L., Tuff, S.: Clin. Chem. *27,* 52 (1981)

Separation Techniques: 124–134

124. Creighton, W. D., Lambert, P. H., Miescher, P. A.: J. Immunol. *3,* 1219 (1973)
125. Ansari, A. A., Bahuguna, L. M., Malling, H. V.: J. Immunol. Methods *26,* 203 (1979)
126. Peterson, M. S., Swerdoff, R. S.: Clin. Chem. *25,* 1239 (1979)
127. Hadassa, H., Berger, S., Bar-Ilan, A., Jonas, Z.: Clin. Chem. *26,* 666 (1980)
128. Remes, V., Slaninova, J., Skopkova, J., Barth, T., Franek, F.: Radiochem. Radioanal. Letters *34,* 253 (1978)
129. Chard, T., in: Methods in Enzymology: Radioimmunoassays and Immunoradiometric Assays, p. 280, New York, Academic Press 1980
130. Jacobs, P. M.: Lig. Quart. *4,* 23 (1981)
131. Bolton, A. E.: Irish Vet. J. *35,* 187 (1981)
132. Cleveland, P. H., Wickham, M. G., Goldbaum, M. H., Ryan, A. F., Worthen, D. M.: J. Immunoassay *2,* 117 (1981)
133. Hersh, L. S., Yaverbaum, S.: Clinica Chimica Acta *63,* 69 (1975)
134. Brown, T. R., Bagchi, N., Ho, T. T. S., Mack, R. E.: Clin. Chem. *26,* 503 (1980)

Radioimmunoassay: 135–143

135. Bolton, A. E.: Irish Vet. J. *36,* 27 (1982)
136. Parker, C. W.: Ann. Rev. Pharmacol. Toxicol. *21,* 113 (1981)
137. Bolton, A. E.: Irish Vet. J. *35,* 119 (1981)
138. Yalow, R. S.: Trends Anal. Chem. *1,* 128 (1982)
139. Berson, S. A., Yalow, R. S.: Clinica Chimica Acta *22,* 51 (1968)
140. Yalow, R. S.: Medical Physics *5,* 247 (1978)
141. Skelley, D. S., Brown, L. P., Besch, P. K.: Clin. Chem. *19,* 146 (1973)
142. Goldsmith, S. J.: Semin. Nucl. Med. *5,* 125 (1975)
143. Dwenger, A.: J. Clin. Chem. Clin. Biochem. *22,* 883 (1984)

Enzymoimmunoassay: 144–150

144. Rubenstein, K. E., Schneider, R. S., Ullman, E. F.: Biochem. Biophys. Res. Commun. *47,* 846 (1972)
145. Litman, D. J., Hanlon, T. M., Ullman, E. F.: Anal. Biochem. *106,* 223 (1980)
146. Howell, E. E., Nasser, J., Schray, K. J.: J. Immunoassay *2,* 205 (1981)
147. Voller, A.: Heterogeneous enzyme immunoassays and their applications, in: Enzyme-Immunoassay, Maggio, E. T. (ed.), p. 181, Boca Raton, CRC Press 1980
148. Landon, J.: Nature *268,* 483 (1977)
149. Carter, J. H.: J. Clin. Immunoassay *7,* 64 (1984)
150. Oellerich, M.: J. Clin. Chem. Biochem. *22,* 895 (1984)

Fluoroimmunoassay: 151–156

151. Cobb, M., Gotcher, S.: Focus *48,* 671 (1982)
152. Gerson, B.: J. Clin. Immunoassay *7,* 73 (1984)
153. Al-Ansari, A. A. K., Smith, D. S., Landon, J.: J. Steroid Biochem. *19,* 1475 (1983)
154. Jolley, M. E., Wang, C.-H. J., Ekenberg, S. J., Zuelke, M. S., Kelso, D. M.: J. Immunol. Methods *67,* 21 (1984)
155. Hicks, J. M.: Human Pathology *15,* 112 (1984)
156. Hemmila, I.: Clin. Chem. *31,* 359 (1985)

Chemiluminoassay: 157–162

157. Gorus, F., Schram, E.: Laboratory Management, September, 45 (1980)
158. Stanley, P. E.: Trends in Anal. Chem. *2,* 248 (1983)
159. Seitz, W. R.: Clin. Biochem. *17,* 120 (1984)
160. Boeckx, R. L.: Hum. Pathol. *15,* 104 (1984)
161. Kricka, L. J., Whitehead, T. P.: Diag. Med., May, 45 (1984)
162. Wood, W. G.: J. Clin. Chem. Clin. Biochem. *22,* 905 (1984)

Other Types of Immunoassay Systems: 163–165

163. Gibbons, I., Hanlon, T. M., Skold, C. N., Russell, M. E., Ullman, E. F.: Clin. Chem. *27,* 1602 (1981)
164. Tyhach, R. J., Rupchock, P. A., Pendergrass, J. H., Skjold, A. C., Smith, P. J., Johnson, R. D., Albarella, J. P., Profitt, J. A.: Clin. Chem. *27,* 1499 (1981)
165. Cais, M.: Metalloimmunoassay: Principles and Practice, in: Methods in Enzymology, p. 445, New York, Academic Press 1983

Immunoassay Method Selection: 166–168

166. Perlstein, M. T.: Lig. Quart. *1,* 4 (1978)
167. Perlstein, M. T.: Lig. Quart. *2,* 6 (1979)
168. Nickoloff, E. L.: in: Nuclear Medicine In Vitro, 2nd Ed., Rothfeld, B. (ed.), p. 35, Philadelphia, J. B. Lippincott 1983

Problems in Immunoassay: 169–223

169. Wertz, P. G., Street, T. L.: Clin. Chem. *29,* 724 (1983)
170. Aldwin, L., Kabakoff, D. S.: Clin. Chem. *27,* 770 (1981)
171. Alexander, N. M., Gatira, R., Nishimoto, M.: Clin. Chim. Acta *100,* 301 (1980)
172. Eriksen, P. B., Anderson, O.: Clin. Chem. *25,* 169 (1979)
173. Ying, C., Miyada, D., Gwinup, G., Simpkins, H.: Clin. Chem. *27,* 344 (1981)
174. Schwertner, H. A., Weintraub, S. T., Hawthorne, S. B.: Clin. Chem. *26,* 1927 (1980)
175. Nickoloff, E. L.: Lig. Quart. 2, 22 (1979)
176. Carmel, R.: Am. J. Clin. Path. *70,* 364 (1978)
177. Witherspoon, L. R., Shuler, S. E., Garcia, M. M., Zollinger, L. A.: Clin. Chem. *25,* 1975 (1979)
178. Kubasik, N. P., Ricotta, M., Hunter, T., Sine, H. E.: Clin. Chem. *28,* 164 (1982)
179. Bauman, J. E.: Clin. Chem. *28,* 2336 (1982)
180. McKiel, R. R., Barron, N., Needham, C. J., Wilkins, T. A.: Clin. Chem. *28,* 2333 (1982)
181. Shah, V. P., Knapp, G., Skelly, J. P., Cabana, B. E.: Clin. Chem. *28,* 2327 (1982)
182. Koida, M., Nakamuta, A., Matsuo, J., Yamamoto, Y., Orlowski, R. C.: Jpn. J. Pharmacol. *32,* 413 (1982)
183. Margry, R. J. C. F., Jacobs, C. W. M., DeGrip, W. J., Daemen, F. J. M.: Vision Res. *22,* 1447 (1982)
184. Furlanetto, R. W.: J. Clin. Endocrinol. Metab. *54,* 1084 (1982)
185. Hammarstrom, M., Mullins, R., Sgoutas, D.: Clin. Chem. *29,* 1418 (1983)
186. von Stetten, O., Schlett, R.: Clin. Chem. *29,* 394 (1983)
187. Malini, P. L., Marata, A. M., Strocchi, E.: Clin. Chem. 28 (1982)
188. Thomas, R. W., Maddox, R. R.: Ther. Drug. Monit. *3,* 117 (1981)
189. Silber, B., Scheiner, L. B., Powers, J. L.: Clin. Chem. 25 (1979)
190. Osterloh, J., Herold, S., Pond, S.: J. Am. Med. Assoc. *247,* 1596 (1982)
191. Valdes, Jr., R., Graves, S. W., Brown, B. A., Landt, M.: J. Pediatr. *102,* 947 (1983)
192. Schwarz, S., Boyd, J.: J. Steroid Biochem. *16,* 823 (1982)
193. Haning, Jr., R. V., Carlson, I. H., Cortes, J., Nolten, W. E., Meier, S.: Clin. Chem. *28,* 696 (1982)
194. Schwartz, H. L., Schadlow, A. R., Faierman, D., Surks, M. I., Oppenheimer, J. H.: J. Clin. Endocrinol. Metab. *36,* 598 (1973)
195. Lundberg, P.-A., Jagenburg, R., Lindstedt, G., Nystrom, E.: Clin. Chem. *28,* 1241 (1982)
196. Saeed-Uz-Zafar, M., Miller, J. M., Breneman, G. M., Mansour, J.: J. Clin. Endocr. *32,* 633 (1971)
197. Riemersma, R. A., Russel, D. C., Oliver, M. F.: Lancet *2,* 471 (1981)
198. Wilkins, T. A., Midgley, J. E. M., Giles, A. F.: Clin. Chem. *28,* 2441 (1982)
199. Wu, J. T.: Clin. Chim. Acta *130,* 47 (1983)
200. Gambino, R.: Lab. Rep. Phys. *5,* 37 (1983)
201. Chopra, I. J., Solomon, D. H.: Int. Encycl. Pharmacol. Ther. *101,* Chap. 4 (1979)
202. Burgi, H., Wimpfheimer, C., Burger, A., Zaunbauer, W., Rosler, H., Lemarchand-Beraud, T.: J. Clin. Endocrinol. Metab. *43,* 1203 (1976)
203. Shaw, W., Hubert, I. L., Spierto, F. W.: Clin. Chem. *22,* 673 (1976)
204. Rash, J. M., Jerkunica, I., Sgoutas, D.: Clin. Chim. Acta *93,* 283 (1979)
205. Jawad, M. J., Wilson, E. A., Rayburn, W. F.: Clin. Chem. *27,* 280 (1981)
206. Rash, J. M., Jerkonica, J., Sgoutas, D. S.: Clin. Chem. *26,* 84 (1980)
207. Fritz, T. J., Bunker, D. M.: Clin. Chem. *28,* 2325 (1982)
208. Hsu, T. H., Nickoloff, E. L., Rock, R. C., Herner, A.: J. Nucl. Med. *19,* 703 (1978)

209. Nickoloff, E. L., Hsu, T. H., Rock, R. C., Perlstein, M.: Abstracts, Clinical Radioassay Society, 4th National Meeting, Los Angeles 1978
210. Vidnes, A., Silsand, T., Ljosland, R., Kapelrud, H., Prydz, P. E.: Scand. J. Clin. Lab. Invest. *42*, 431 (1982)
211. Allan, D. J., Murphy, F., Needham, C. A., Barron, N., Wilkins, T. A., Midgley, J. E. M.: Lancet *2*, 824 (1982)
212. Bhagat, C. I., Garcia-Webb, P., Watson, F., Beilby, J. P.: Clin. Chem. *29*, 1324 (1983)
213. von Schenck, H., Grubb, A. O.: Clin. Chem. *28*, 1103 (1982)
214. Vladutiu, A. O., Sulewski, J. M., Pudlak, K. A., Stull, C. G.: J. Am. Med. Assoc. *248*, 2489 (1982)
215. Hunter, W. M., Budd, P. S.: Lancet *2*, 1136 (1980)
216. Howanitz, P. J., Howanitz, J. H., Lamberson, H. V., Benedict, S.: Clin. Chem. *26*, 1020 (1980)
217. Spencer, C. A., Challand, G. S.: Clin. Chem. *28*, 584 (1977)
218. Holtzman, J. L., Shafer, R. B., Erickson, R. R.: Clin. Chem. *20*, 1194 (1974)
219. Voshall, D. L., Hunter, L., Grady, H. J.: Clin. Chem. *21*, 402 (1975)
220. Seino, Y., Matsukura, S., Miyamoto, Y., Goto, Y., Taminato, T., Imura, H.: J. Clin. Endocrinol. Metab. *43*, 852 (1976)
221. Lipson, A., Hsu, T.-H., Nickoloff, E. L.: J. Clin. Endocrinol. Metab. *50*, 176 (1980)
222. Molin, L., Bergdahl, B.: Clin. Chem. *29*, 734 (1983)
223. Ghielmi, S., Pizzoccolo, G., Iacobello, C., Albertini, A., Arosio, P.: Clin. Chim. Acta *120*, 285 (1982)

Quality Control: 224–235

224. Dixon, K., Northam, B. E.: Clin. Chim. Acta *30*, 453 (1970)
225. Jeffcoate, S. L., Rose, E. G., Das, M.: Ann. Clin. Biochem. *14*, 258 (1977)
226. Jeffcoate, S. L.: Ann. Clin. Biochem. *18*, 1 (1981)
227. Broughton, P. M. G.: Ann. Clin. Biochem. *6*, 109 (1969)
228. Jeffcoate, S. L.: J. Nucl. Med. Allied Sci. *26*, 249 (1982)
229. Goh, H. H.: Singapore J. Obst. Gynaecol. *14*, 17 (1983)
230. Woodrum, C. E.: MLO, August, 99 (1983)
231. Miller, C. E., Besch, P. K., Hannon, W. H., Knight, W. S.: NCCLS Federal Register 41 (1985)
232. Westgard, J. O., Barry, P. L., Hunt, M. R.: Clin. Chem. *27*, 493 (1981)
233. Hoffman, R. G., Waid, M. E.: Am. J. Clin. Pathol. *43*, 134 (1965)
234. Tonks, D. B.: Clin. Chem. *9*, 127 (1963)
235. Levey, S., Jennings, E. R.: Am. J. Clin. Pathol. *20*, 1059 (1950)

Normal Range, Reference Range: 236–246

236. Galen, R. S.: Arch. Pathol. Lab. Med. *101*, 561 (1977)
237. Grasbeck, R., Saris, N. E.: Scand. J. Clin. Lab. Invest. *24*, 62 (1969)
238. Grasbeck, R., Siest, G., Wilding, P.: Clin. Chim. Acta *87*, 461F (1978)
239. Ash, K. O.: Am. J. Med. Tech. *46*, 504 (1980)
240. Amador, E., Hsi, B. P.: Am. J. Clin. Pathol. *52*, 538 (1969)
241. Mainland, D.: Clin. Chem. *17*, 267 (1971)
242. Reed, A. H., Henry, R. J., Mason, W. B.: Clin. Chem. *17*, 275 (1971)
243. Herrera, L.: J. Lab. Clin. Med. *52*, 34 (1958)
244. Sunderman, F. W., Jr.: Clin. Chem. *21*, 1873 (1975)
245. Dybkaer, R.: Scand. J. Clin. Lab. Invest. *29* (Suppl. 126), 19.1 (1972)
246. Taylor, W. F.: Scand. J. Clin. Lab. Invest. *29* (Suppl. 126), 19.12 (1972)

Cyclosporine: Chemistry, Structure-Activity Relationships and Mode of Action

Roland M. Wenger, Trevor G. Payne and Max H. Schreier

Preclinical Research, Sandoz Ltd, CH-4002 Basel, Switzerland

Cyclosporine is a cyclic peptide with a selective immunosuppressive action. One of its 11 amino acids is a novel amino acid, (4*R*)-4-((*E*)-2-butenyl-4,N-dimethyl-L-threonine, the preparation of which allowed the synthesis of cyclosporine and specifically modified analogues. Structure-activity relationships emerging from the study of modified cyclosporines suggest that a large portion of the cyclosporine molecule is involved in interactions with its lymphocyte receptor and that this includes amino acids 1, 2, 3 and 11.

Cyclosporine acts mainly on T cells and affects the induction phase rather than the proliferative phase of lymphoid populations. Its primary action appears to be the inhibition of a number of lymphokines such as IL-2, gamma-IFN and OAF at the level of transcription or at an earlier activation step which selectively leads to transcription of these and not other proteins. Although a cytosolic receptor protein and a number of sub-cellular events affected by cyclosporine have been demonstrated, the precise molecular mechanism remains to be identified.

Progress in Clinical Biochemistry
and Medicine, Vol. 3

1 Introduction

The cyclic undecapeptide cyclosporine[1] is unique among the presently available immunosuppresive drugs in that it reversibly inhibits only some classes of lymphocytes and does not affect haemopoetic tissues. Initially isolated for its antifungal activities, it was characterized by Borel et al. as an immunosuppressant which acted mainly on T-lymphocytes and showed antiinflammatory effects[2–6]. More recently its spectrum of activity has been shown to include a number of antiparasitic actions. Bueding et al.[7] found that the compound had antischistosomal activity and Thommen[8] was the first to demonstrate its antimalarial action, both activities probably being unrelated to immunosuppression.

Cyclosporine has been reported to be effective in preventing allograft rejections for a variety of organs in a number of different species (for reviews see[9, 10]). The first transplantations in humans performed with the aid of cyclosporine were reported by Calne et al.[11] for kidney, and by Powles et al.[12] for bone marrow. Since then the use of cyclosporine has been extended to include liver, pancreas, lung, and corneal transplantations and the availability of the drug has led to a revival in heart transplantations. Cyclosporine was introduced on the market in 1983 for the prevention of organ rejection under the trade name SANDIMMUNE® and its use in this indication has been the subject of a number of reviews[13]. The extension of the use of cyclosporine to other clinical indications such as autoimmune diseases is currently under investigation.

In the present review only three aspects of cyclosporine will be presented. In the first section on the chemistry of cyclosporine, the main emphasis will be on describing the unique structural features of the molecule and its synthesis. A brief discussion of its biosynthesis will be presented for comparison. In the second section, the results obtained from specifically modified synthetic cyclosporines and some naturally occurring variants are interpreted in terms of the structural requirements for immunosuppressive activity. Although incomplete, these results are beginning to define a large area on the surface of the cyclosporine structure which must be involved in interactions between this compound and its receptor on or in the lymphocyte. The detailed structure-activity relationships strongly support a discrete interaction with a target protein rather than a general alteration of membrane properties. In the third section some of the effects of cyclosporine in lymphoid cell culture systems are discussed in terms of its possible mechanism of action.

1.1 Production of Cyclosporine

Cyclosporine[1], originally named "cyclosporin A"[14], is produced by fungi of the genus Tolypocladium in submerged culture as described by Dreyfuss et al.[15] and under normal fermentation conditions it is the main component of a group of cyclic undecapeptides. The naturally occurring cyclosporine analogues, cyclosporines A to I, have been reported by Traber et al.[16, 17] and using extensive chromatographic procedures a further 16 minor metabolites, cyclosporines I to Z have been isolated and identified out of fermentations of Tolypocladium inflatum Gams[18]. In many of the metabolites methylation of the amide linkages has not proceeded to the same extent as in cyclosporine and others differ in their structure at one or two amino acids. The most frequent variation is at position 2 where L-α-aminobutyric acid can be replaced by L-alanine, L-threonine, L-valine and L-norvaline. In position 1, N-methyl-L-leucine, N-methyl-L-α-amino-octanoic acid and desoxy-MeBmt have been found. Other variations include L-norvaline in position 5, L-α-aminobutyric acid in position 7 and D-N-methylvaline in position 11. In some cases the biosynthesis of the cyclosporines can be directed by externally supplying the corresponding precursor. For example (norvaline^2)cyclosporine can be selectively produced by adding L-norvaline to the fermentation broths[19]. Addition of D,L-α-aminobutyric acid increased production of cyclosporine and almost completely suppressed the formation of cyclosporine analogues.

1.2 Elucidation of the Structure of Cyclosporine

The cyclosporine structure (Fig. 1) was established by chemical degradation[14] together with an X-ray crystallographic analysis of an iodo derivative (**a**). This analysis, as described by Petcher et al.[20], not only determined the sequence of the native peptide, but also gave an insight into its shape. The structural and conformational assignments have later been confirmed[21] by an X-ray crystallographic analysis and by two-dimensional nuclear magnetic resonance (NMR) studies of cyclosporine itself.

Cyclosporine (Fig. 1) is a neutral, hydrophobic, cyclic peptide composed of eleven amino acid residues, each having the *S*-configuration of the natural L-amino acids, except for the D-alanine in position 8, which has the *R*-configuration and sarcosine (N-methyl-glycine) in position 3. The seven amino acids in positions 1, 3, 4, 6, 9, 10 and 11 are N-methylated thereby restricting the number of possibilities for intramolecular hydrogen bonds. Except for the amino acid in position 1, all other amino acids were known aliphatic amino acids and were readily characterized following acid hydrolysis of cyclosporine. The novel amino acid in position 1, previously referred to as the "C-9 amino acid"[14], being composed of 9 carbon atoms, is now designated as (4*R*)-4-[(*E*)-2-butenyl]-4,N-dimethyl-L-threonine and abbreviated MeBmt[22]. Thus the novel amino acid has the polar features of an N-methyl-L-threonine which is substituted at the end of the carbon chain by butenyl and methyl groups. This amino acid was hitherto unknown in its free form as only artifacts and

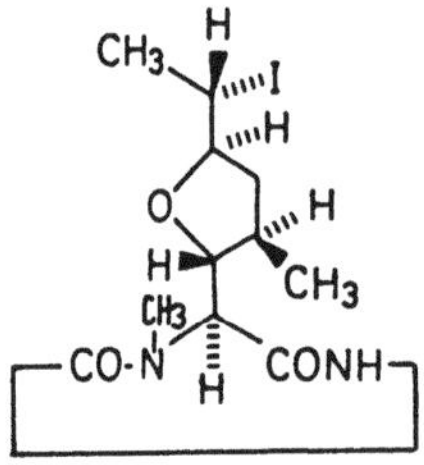

a

a) Iododerivative of cyclosporine (partial structure)

derivatives were obtained in the course of cyclosporine degradation experiments[14]. Hydrolysis of cyclosporine followed by ion-exchange chromatography produced a cyclic derivative (**b**) of the MeBmt amino acid as the only isolable MeBmt-derivative. Acidic treatment of cyclosporine (**c**) in the absence of water effected an N,O-carbonyl migration of the MeVal-moiety and gave isocyclosporine (**d**) which with methylisothiocyanate under the conditions of a modified Edman degradation pro-

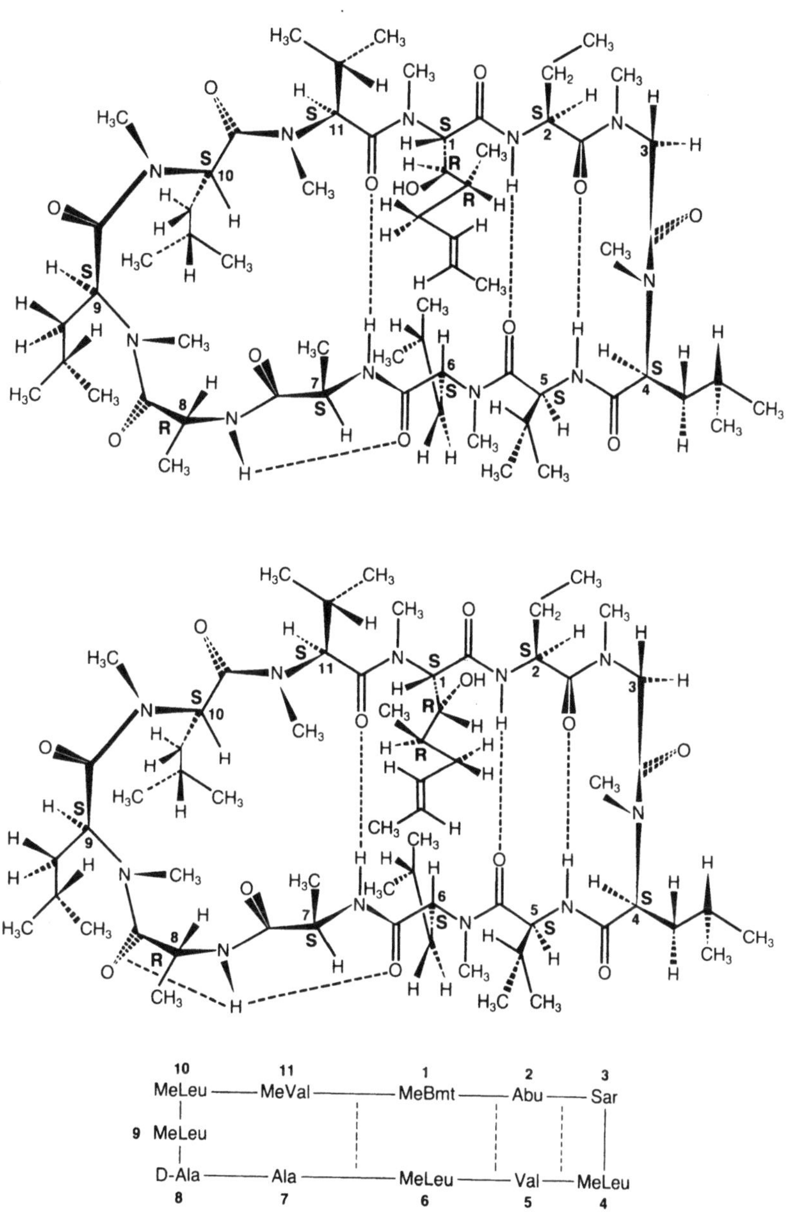

Fig. 1. Schematic representation of the structure and conformation of cyclosporine in the solid state (above) and in solution (below, apolar solvents)

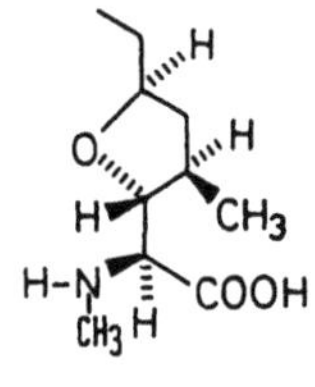

b

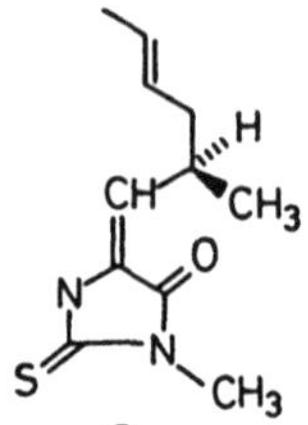

e

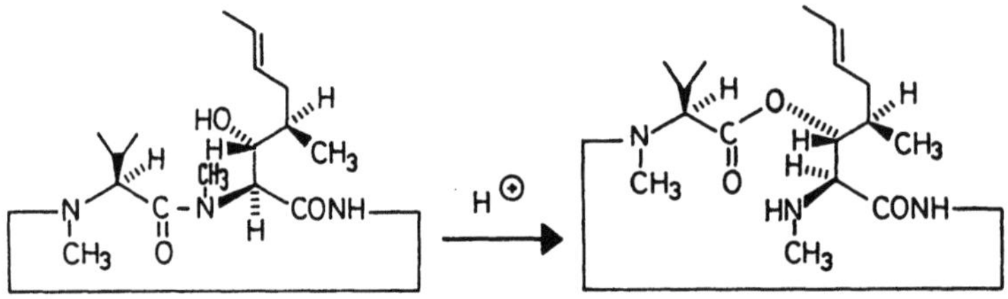

c d

b) Cyclic Derivative of MeBmt; **e**) Anhydromethylthiohydantoin Derivative of MeBmt; **c**) Cyclosporine (Partial Structure); **d**) Isocyclosporine (Partial Structure)

duced an anhydrothiohydantoin-derivative of MeBmt (**e**). This reaction established the MeBmt as the first amino acid of the sequence and permitted the complete sequence elucidation by repetitive Edman degradation.

The X-ray structure determinations and NMR measurements indicate that cyclosporine assumes a rather rigid conformation both in the crystalline state and in solution. A large portion of the backbone, residues 1 to 6, adopt an antiparallel beta-pleated sheet conformation which contains three transannular hydrogen bonds and is markedly twisted. The sarcosine in position 3 and the N-methylleucine in position 4 participate in a type II′ beta-turn[23–25)] which means that, as drawn in Fig. 1, the C=0 of Abu-2 and the N-methyl of MeLeu-4 are up and the C=0 of Sar-3 and NH of Val-5 are down, the pro-*S* proton of the methylene group of Sar-3 is axial and the carbon chain of MeLeu-4 equatorial to the peptide ring. The remaining residues 7 to 11 form an open loop which contains the only cis amide linkage in the molecule between the two adjacent N-methylleucine residues 9 and 10. The remaining H-bond observed in the crystalline state is of a gamma-type and serves to hold the backbone in a folded L shape. A minor difference in the backbone conformation in the region of D-Ala-8 is observed in chloroform solution as a three-centre hydrogen bond has been postulated to occur between the NH of D-Ala-8 and the carbonyl oxygens of both Leu-6 and D-Ala-8[21)]. The major difference between the crystal and the solution conformations is the orientation of the carbon chain of the MeBmt amino acid (χ_1 = −168° in the crystal, +60° in solution) due to the formation of a hydrogen bond between the hydroxyl and the carbonyl of MeBmt in the latter case.

As a consequence of the rather rigid conformation of the cyclosporine skeleton, 6 amino acids have their carbon chains directed quasi perpendicular to the plane of the peptide ring. They are Abu-2, Val-5, Ala-7 and MeVal-11, projecting up and MeBmt-1 and MeLeu-6 projecting down in Fig. 1. The remaining 5 amino acids have their carbon chain more or less in the plane of the peptide ring.

1.3 Biosynthesis of Cyclosporine

In the initial biosynthetic studies of Kobel et al.[26], ^{3}H and ^{13}C labelled precursors were fed to the culture and the position of incorporation of the label in cyclosporine determined by NMR spectroscopy. It was demonstrated that the N-methyl groups of cyclosporine and the methyl group in the gamma-position of the unsaturated amino acid MeBmt are introduced as intact methyl groups from methionine and that the remaining carbons of the MeBmt moiety are derived from the head to tail condensation of 4 acetate units. The ^{13}C-NMR spectrum of the enriched cyclosporine derived from [1-^{13}C]-acetate showed 4 enhanced signals corresponding to the carbon atoms 1, 3, 5 and 7 of the Bmt unit and no ^{13}C incorporation into any other amino acid.

In an analogous example of non-ribosomal cyclic peptide synthesis by fungi, Zocher and Kleinkauf[27] showed that in the case of enniatin a mulitifunctional enzyme carries out the N-methylation of the constituent amino acids subsequent to the ATP dependent activation step when they are bound to the enzyme as thioesters. In similar studies on cyclosporine biosynthesis[28] support for a similar mechanism was implied from the observation that ^{14}C-sarcosine was not incorporated and that the radioactivity from [Me-^{14}C] methionine incorporated into each N-methylated amino acid was directly proportional to the number of residues in cyclosporine indicating that N-methylation of the amino acids occurred simultaneously.

From the results of Kobel et al.[26] the site of biosynthesis of the Bmt unit remains unclear. However, with short-term feeding experiments using labelled acetate and methionine, the Kleinkauf group[28] only detected incorporation of label into the N-methyl groups suggesting that the amino acid Bmt is not biosynthesized on the enzyme matrix but at some other site before incorporation into cyclosporine.

1.4 Pharmokinetics and Metabolism of Cyclosporine

The studies of Wood et al.[29] showed that after oral administration of the olive oil-based drinking solution, the bioavailability of cyclosporine is 20 to 50%. Peak blood concentrations are usually reached after 3–4 h. Cyclosporine is extensively distributed to extravascular tissues because of its high lipid solubility. In the blood, approximately 50% of the drug is associated with erythrocytes, 10–20% with leukocytes, and of the 30–40% found in plasma, about 90% is bound to lipoproteins. The drug is eliminated primarily by biliary mechanisms: the terminal elimination half life being in the range of 14 to 27 h. Approximately 17 metabolites are formed from the parent drug, all of which are present in considerably lower concentrations than cyclosporine itself. The metabolism of cyclosporine has been studied in the rat, rabbit, dog and man[30] and the biotransformation pathways are similar in all species. Nine of the metabolites have been isolated and identified and all retained the intact cyclic peptide structure of the parent drug. Structural modifications originated from oxidation of only 4 of the 11 amino acids. Hydroxylation reactions appeared to be restricted to the terminal carbon atom (8′) of amino acid 1 (MeBmt) and the gamma-position (4′) of the N-methylleucines 4, 6 and 9. Oxidative N-demethylation, which presumably also proceeds via hydroxylation, appears to only occur on N-methylleucine-4. The primary metabolites are the monohydroxylated cyclosporines (gamma-hydroxy-MeLeu^{9}) cyclosporine, (8'-hydroxy-MeBmt^{1})cyclosporine, and the N-demethylated (Leu^{4})cyclosporine. The monohydroxylated derivatives can serve as substrates for further metabolism, hydroxylation then occurring on amino acids 1, 4 or 6. The identification of the metabolite (gamma-hydroxy-MeLeu^{6,9},Leu^{4})cyclosporine indicates either that the N-demethylation is possible on the metabolite hydroxylated on the gamma-carbon of leucines 4 and 6, or that the primary N-demethylated metabolite (Leu^{4})cyclosporine can be oxidized by the enzymes hydroxylating N-methylleucines 6 and 9. The structure of several further urinary metabolites remained incompletely determined.

2 Synthesis of Cyclosporine

2.1 Synthesis of the Enantiomerically Pure Amino Acid Bmt[31)]

The basic problems associated with the synthesis of (4*R*)-4-((*E*)-2-butenyl-4,N-dimethyl-L-threonine (MeBmt) are those of constructing three contiguous asymmetric centres with the correct stereochemical configuration and ensuring that only the *trans* isomer of the double bond is formed. Using the Caln-Ingold-Prelog rules[32)] to specify stereochemistry, the amino acid can be designated as (2*S*,3*R*,4*R*,6*E*)-3-hydroxy-4-methyl-2-methylamino-6-octenoic acid.

Tartaric acid has been established as a practical source of chiral building blocks for synthesis[33)] and (*R*,*R*)-(+)-tartaric acid was used as the starting material for this synthesis. It was modified in three major operations to introduce the features of the target amino acid. In the first operation, summarized in Scheme 1, one OH-group of the (*R*,*R*)-(+)-tartaric acid molecule was incorporated with the correct configuration and the other OH-group was replaced by a methyl group with inversion of

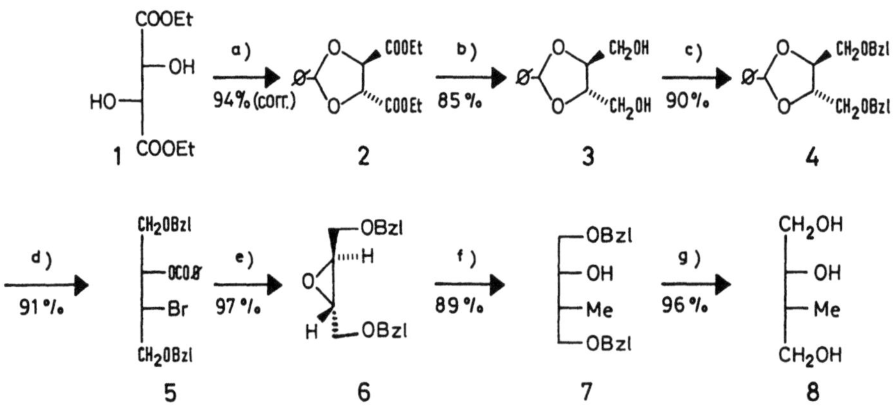

1 2 3 4 5 6 7 8

Scheme 1. Synthesis of (R,R)-3-Methyl-1,2,4-butanetriol (**8**). a) PhCHO/HC(OEt)$_3$/TsOH. b) LiAlH$_4$. c) BzlBr/KOH. d) NBS. e) KOH/EtOH. f) 2 MeLi/CuI. g) Pd/H$_2$

configuration. This provided the asymmetric centres C(3) and C(4) of the amino acid MeBmt. The second operation consisted of introducing the (*E*)-butenyl moiety (Scheme 2) after having elongated the carbon chain by formation of an aldehyde which was the starting material for a modified Wittig reaction. The third operation consisted of oxidizing the diol **14** to the hydroxy aldehyde **19** (Scheme 3) and intro-

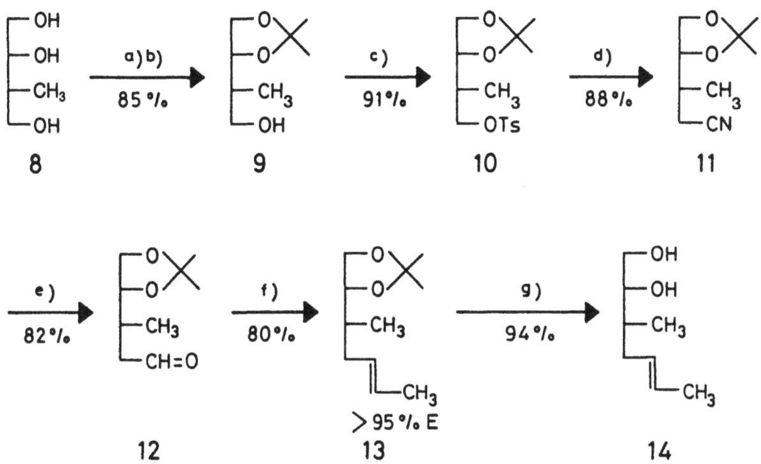

Scheme 2. Synthesis of (R,R,E)-3-methyl-5-heptene-1,2-diol (**14**). a) $Me_2C(OMe)_2/TsOH/C_6H_6$, reflux, 2 h. b) Acetone, TsOH, reflux, 15 h. c) TsCl/Py, 35°, 4 h. d) KCN/DMSO, 20°, 3 days. e) DIBAH/hexane, –75°, 2 h. f) Ph_3EtPBr/BuLi, Schlosser conditions. g) 1.1 equiv. of 1 N HCl, THF/H_2O 4:1, 20°, 2 days

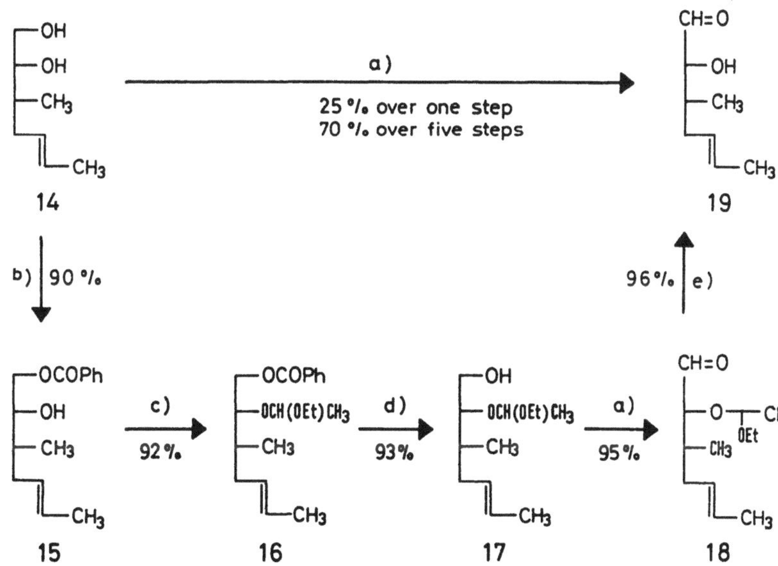

Scheme 3. Synthesis of (R,R,E)-2-Hydroxy-3-methyl-5-heptenal (**19**). a) $DCCI/DMSO/C_6H_6/Py/TFA$, 20°, 2 h. b) PhCOCl/Py, 20°, 1 h. c) $CH_2{=}CHOEt/TFA$, 20°, 1–3 days. d) 10 N KOH/EtOH, 20°, 1.5 h. e) 1 N HCl/THF, 20°, 2 h

ducing the MeNH- and the COOH groups via a cyclic intermediate permitting stereochemical control during the formation of the third asymmetric centre at C(2) (Scheme 4).

2.1.1 Synthesis of (R,R)-3-Methyl-1,2,4-Butanetriol (**8**)*. Formation of the Asymmetric Centres for C(3) and C(4) (Scheme 1)*

Diethyl tartrate (**1**) was treated with benzaldehyde in the presence of triethyl orthoformate and p-toluene sulfonic acid (TsOH) to form the acetal **2** and thereby protect the OH-groups. Subsequent reduction with $LiAlH_4$ in tetrahydrofuran (THF) to the diol **3** followed by benzylation furnished the dioxolane **4**. In order to transform one of the OH groups into a methyl group with inversion of configuration, the intermediatory epoxide **6** was prepared using a procedure reported by Seeley and McElwee[34] for the conversion of acetals of 1,2 diols into epoxides. Due to the symmetry of **4** it was immaterial on which of the two secondary carbon atoms the displacement occurred. Thus, treatment of **4** with N-bromosuccinimide (NBS) in carbon tetrachloride produced the bromo ester **5** which was directly converted to the optically active and therefore trans-disubstituted oxirane **6** during alkaline hydrolysis. The alkylation of the symmetrical epoxide was achieved using the method of Johnson et al.[35] with excess MeLi(CuI) in ether at –15° giving a single product **7**. Hydrogenolysis in ethanol removed the benzyl protecting groups giving the triol **8** in an overall yield of 56% from diethyl tartrate.

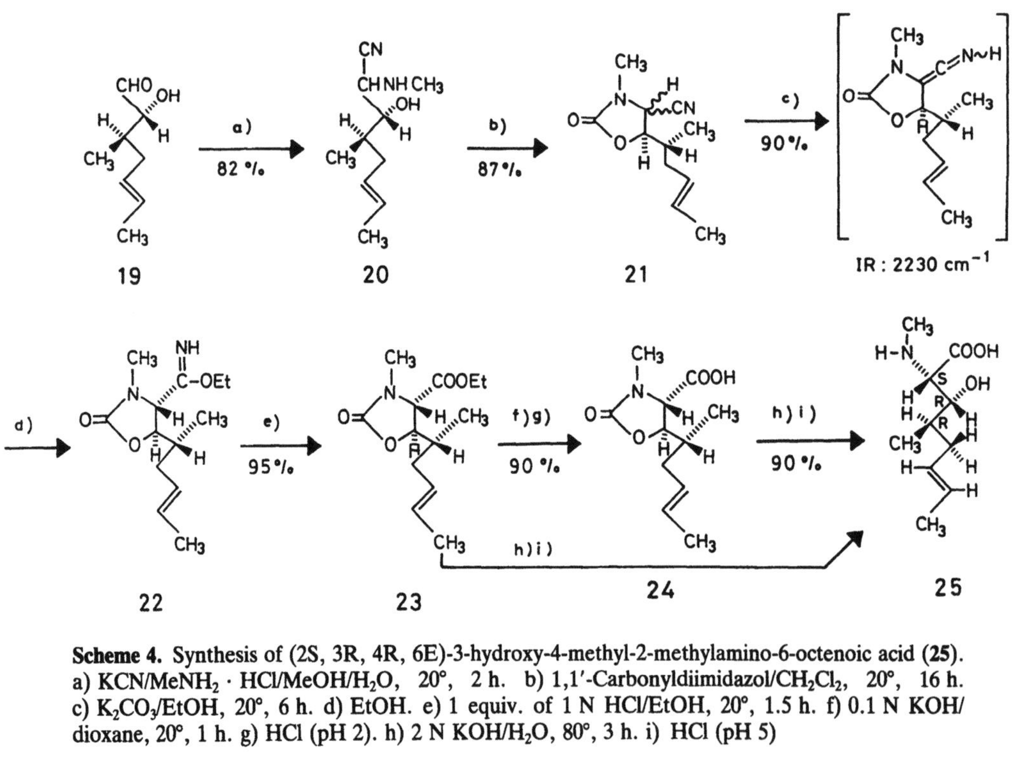

Scheme 4. Synthesis of (2S, 3R, 4R, 6E)-3-hydroxy-4-methyl-2-methylamino-6-octenoic acid (**25**). a) $KCN/MeNH_2 \cdot HCl/MeOH/H_2O$, 20°, 2 h. b) 1,1′-Carbonyldiimidazol/CH_2Cl_2, 20°, 16 h. c) K_2CO_3/EtOH, 20°, 6 h. d) EtOH. e) 1 equiv. of 1 N HCl/EtOH, 20°, 1.5 h. f) 0.1 N KOH/dioxane, 20°, 1 h. g) HCl (pH 2). h) 2 N KOH/H_2O, 80°, 3 h. i) HCl (pH 5)

2.1.2 *Synthesis of (R,R,E)-3-Methyl-5-Heptene-1,2-Diol* (**14**). *Chain Elongation with Introduction of the Trans-Double Bond (Scheme 2)*

The two vicinal OH-groups of the triol **8** were selectively protected as the ketal **9**, which was formed together with 10–15% of the isomeric 1,3-dioxane via an acid catalysed reaction of **8** with 2,2-dimethoxypropane in benzene. To elongate the carbon chain, the dioxolane **9** was converted to the corresponding tosylate **10** and then treated with KCN in dimethylsulfoxide to produce the nitrile **11** which was reduced with diisobutylaluminium hydride at –75° to the aldehyde **12**. This aldehyde was then subjected to a Wittig reaction using ethyl(triphenyl)phosphonium bromide under the conditions recommended by Schlosser and Christmann[36)] in order to ensure the formation of the desired *trans* isomer of the olefin **13**. Treatment of **13** with 1N HCl in aqueous tetrahydrofuran removed the isopropylidene-protecting group and gave the olefinic diol **14** in an overall yield of 42% from **8**.

2.1.3 *Synthesis of (R,R,E)-2-Hydroxy-3-Methyl-5-Heptenal* (**19**) *(Scheme 3)*

The oxidation of the diol **14** to the corresponding hydroxy aldehyde **19** could be achieved in one step using the Pfitzner-Moffatt procedure[37)], but only in low yield (25%). To obtain the hydroxy aldehyde in high yield it is necessary to protect the secondary OH-group of **14**. This was done by monobenzoylation (**15**), followed by protection of the secondary OH-group as the ethoxyethyl derivative **16** (ethylvinyl ether/trifluoroacetic acid), and alkaline hydrolysis of the benzoate to give the primary alcohol **17**. Oxidation of **17** to the aldehyde **18** under the above conditions proceeded with a yield of 95%. Removal of the ethoxyethyl-protecting group with 1N HCl in tetrahydrofuran at room temperature gave the hydroxyaldehyde **19** in a total yield over five steps of 70%.

2.1.4 *Introduction of the Methylamino and the Carboxy Groups. Completion of the Synthesis of MeBmt (25) (Scheme 4)*

Treatment of freshly prepared aldehyde **19** at room temperature with KCN and methylamine hydrochloride in aqueous methanol yielded the cyanamine **20** as a mixture of diastereoisomers. On reacting with 1,1'carbonyldiimidazol, the mixture **20** was converted to the oxazolidine-2-one diastereoisomers **21** (6:1/*cis*:*trans* rel. to ring) which on treatment with potassium carbonate in ethanol formed a single carboximidate **22**. The intermediate didehydroimine (Scheme 4) can be characterized by a band at 2230 cm^{-1} in the IR spectrum of a crude product. This N,α-didehydroimine reacted stereospecifically with ethanol to yield the thermodynamically more stable trans (rel. to ring) carboximidate **22**. Hydrolysis of **22** with 1 equivalent of 1 N HCl gave the enantiomerically pure N-methyl-amino acid derivative **23** with the O- and N-functional groups in the desired *threo*-configuration. Both protecting groups of **23** can be removed in one step by treatment with 2 N KOH at 80° or stepwise by selective hydrolysis of the ester linkage with 0.1 N KOH in dioxane at room temperature to form the acid **24** followed by cleavage of the oxazolidine-2-one ring with 2N

KOH at 80°. On acidifying the reaction mixtures to pH 5, the desired N-methylamino acid **25** crystallized from the reaction mixture and was isolated in a yield of 48% from **19**. The overall yield of **25** from diethyl tartrate after 24 steps was 7.8%, the average yield per step being 90%.

The stereospecific synthesis described above allowed the new amino acid (4*R*)-4-((*E*)-butenyl)-4,N-dimethyl-L-threonine (**25**, MeBmt) to be characterized for the first time and opened the way for a total synthesis of cyclosporine.

2.2 *Strategy Used for the Synthesis of Cyclosporine*[38, 39]

For the synthesis of cyclosporine (Fig. 1), the peptide bond between the L-alanine in position 7 and the D-alanine in position 8 was chosen for the cyclisation step for two main reasons. The intrachain H-bonds between amide groups of the linear peptide may in this case operate so as to stabilize the open chain in folded comformations approximating the cyclic structure of cyclosporine and thus assist cyclisation. The strategy was also specifically chosen to preclude an N-methylamino acid at the N- and C-terminus of the undecapeptide since bond formation between N-methylated amino acids presents more difficulties than for non-N-methylated derivatives[38, 40]. Therefore bond formation between the only two consecutive, non-N-methylated amino acids in cyclosporine appeared the logical choice for the cyclisation step.

For the synthesis of the undecapeptide, a fragment-condensation technique introducing the amino acid MeBmt at the end of the synthesis was used. In this way, the number of steps after the introduction of this amino acid was minimized. The peptide fragments were built up in the direction shown by the arrows in Scheme 5 using the step sequence which is indicated numerically. The amino groups of the amino acids and peptides being reacted were generally protected with a tert-butoxycarbonyl group (Boc) and the carboxy groups with a benzyloxy group (benzyl ester; OBzl). The carboxy groups were normally activated using a variation of the mixed pivalic anhydride method reported by Zaoral[41] and adapted for N-methyl-amino acid derivatives[38] by allowing slow anhydride formation in chloroform at –20° with pivaloylchloride (= trimethylacetylchloride) in the presence of 2 equivalents of a tertiary base such as N-methylmorpholine before adding the amino acid or peptide esters to be coupled as the free base. The Boc protecting groups of the peptide intermediates were removed with trifluoroacetic acid at –20°, the acids neutralized with sodium bicarbonate and the peptides isolated as the free bases. The benzyloxy groups of the peptide intermediates were removed by catalytic hydrogenation in ethanol using a palladium on carbon catalyst.

The tetrapeptide Boc-D-Ala-MeLeu-MeLeu-MeVal-OBzl was made from the left to the right by forming bonds 1, 2 and 3 (Scheme 5). The hexapeptide on the carboxyl side of MeBmt was constructed from the dipeptide Boc-Abu-Sar-OBzl and the tetrapeptide Boc-MeLeu-Val-MeLeu-Ala-OBzl each being synthesized from the left to the right as shown in Scheme 5 thereby minimizing the risk of racemization. Bond 8 was then formed giving the hexapeptide Boc-Abu-Sar-MeLeu-Val-MeLeu-Ala-OBzl which was then deprotected at the amino end. The hydroxy and N-methylamino functions of the amino acid MeBmt were protected in the five membered ring

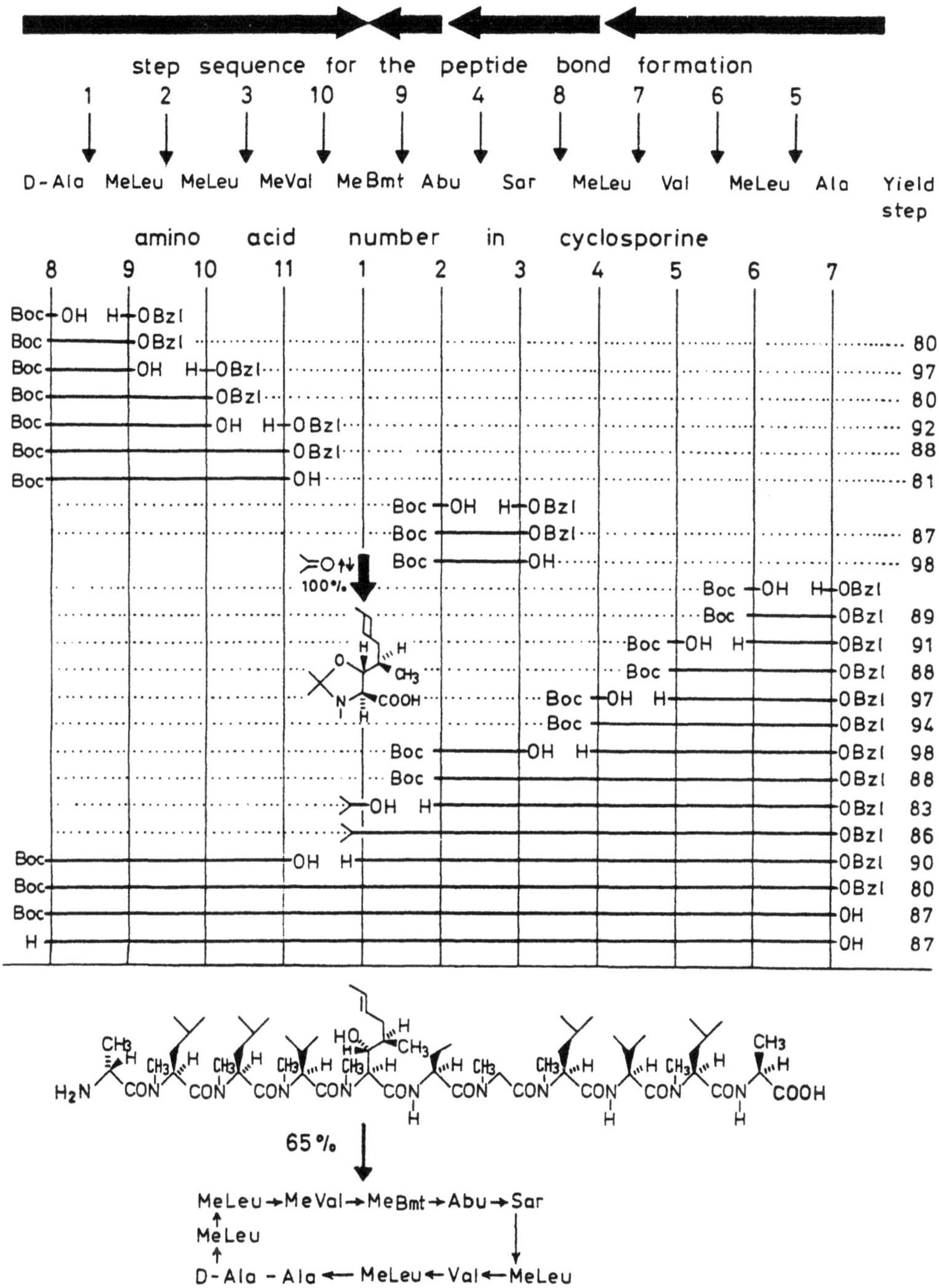

Scheme 5. Synthesis of cyclosporine. (strategy used for the synthesis of the peptides)

of a dimethyloxazolidine (isopropylidene protecting group) which ensured that the thermodynamically more stable trans configuration of the substituents was maintained during carboxy activation and peptide formation. The isopropylidene derivative was readily prepared by refluxing the amino acid in acetone and was stabilized before activation by the addition of 1 equiv. of N-methylmorpholine. To couple the protected MeBmt amino acid to the hexapeptide, dicyclohexylcarbodiimide (DCCI) in the presence of N-hydroxy-benztriazol (BtOH)[42] was used. The isopropylidene protecting group of the heptapeptide was then removed with 1 equiv. of 1 N HCl in methanol to produce H-MeBmt-Abu-Sar-MeLeu-Val-MeLeu-Ala-OBzl. The final amide linkage 10 was made to produce the undecapeptide Boc-D-Ala-MeLeu-MeLeu-MeVal-MeBmt-Abu-Sar-MeLeu-Val-MeLeu-Ala-OBzl by coupling Boc-D-Ala-MeLeu-MeLeu-MeVal-OH with the heptapeptide by using the reagent (1H-benzo[d][1,2,3]triazol-1-yloxy)tris-(dimethylamino)-phosphonium hexafluorophosphate ($BtOP(NMe_2)_3^+PF_6^-$) developed by Castro et al.[43]. The ester group of the undecapeptide was removed by alkaline hydrolysis and the Boc group cleaved with trifluoroacetic acid. Then, the unprotected H-D-Ala-MeLeu-MeLeu-MeVal-MeBmt-Abu-Sar-MeLeu-Val-MeLeu-Ala-OH was cyclised in dilute methylene chloride solution (0.0002 M) with 1 equiv. of $BtOP(NMe)_3^+PF_6^-$ [43] in the presence of N-methylmorpholine at room temperature to give crystalline cyclosporine, isolated in 62% yield. Using the mixed phosphonic anhydride method described by Wissman and Kleiner[44] or the pentafluorophenol-DCCI complex of Kovacs[45] to effect the cyclisation enhanced the yield of cyclosporine to 65%. Thus using MeBmt and the tetra- and hexapeptide fragments described above, it is now possible to efficiently synthesize cyclosporine in a yield of 27.5% with respect to the amino acid MeBmt. This synthesis of cyclosporine has also opened the way for the preparation of analogs needed to attack the many unanswered problems concerning the structure-activity relationships of this drug.

3 Synthetic Analogues of Cyclosporine and Structure-Activity Relationships

3.1 Choice of the Variations

Some naturally occurring modifications of the cyclosporine peptide structure have been described by Ruegger et al.[14] and Traber et al.[16–18] and their contribution to our understanding of structure-activity relationships will be described below. Using the synthetic approach potentially any amino acid of the peptide chain of cyclosporine (Fig. 1) can be modified and specific aspects of the structure-activity relationships examined one after the other. In order to study the importance of the near environment of the MeBmt amino acid in position 1 of cyclosporine, variations at positions 1, 2, 3 and 11 were first undertaken. Eight such variations will be described here together with the synthesis of (N-methyl-D-valine[11])cyclosporine, a naturally occurring isomer of cyclosporine (Fig. 2).

Earlier work had suggested that the unusual amino acid MeBmt may be important for the biological activity of cyclosporine[20]. In order to study the relevance of the alkyl chain of (4*R*)-4-((*E*)-2-butenyl)4,N-dimethyl-L-threonine in position 1 of cyclosporine, this amino acid was replaced by N-methyl-threonine (MeThr) which retains the polar features of MeBmt but not the extended lipophilic carbon chain.

Another question considered was the possible role of the threonine hydroxy group of (threonine²)cyclosporine, a naturally occurring ring isomer of cyclosporine and potent immunosuppressant. To examine whether or not this hydroxy group was involved in some special receptor interaction, a modified cyclosporine with serine in position 2 was prepared.

The three dimensional structure of cyclosporine as deduced by X-ray crystallographic analysis shows sarcosine in position 3 participating in a type II′ beta-turn (see Sect. 1.2). Replacement of sarcosine by D-proline would be expected to stabilize this structural feature and make it more rigid. The proline ring would be expected to take an equatorial position relative to the peptide ring and should not alter the conformation of the peptide backbone relative to that of cyclosporine. For comparison purposes the corresponding compound with L-proline in position 3, a variation expected to destabilize the type II′ beta-turn, was considered of value to confirm the conformational hypothesis for the (D-Pro³)cyclosporine.

Replacing MeVal-11 in cyclosporine by MeLeu would be expected to relieve steric crowding and allow more flexibility in the peptide ring. The MeVal at this position was replaced by MeLeu, an amino acid having one additional methylene group in its carbon chain. This variation also introduces additional steric bulk and the sensitivity of the biological activity to minor structural changes in this region of

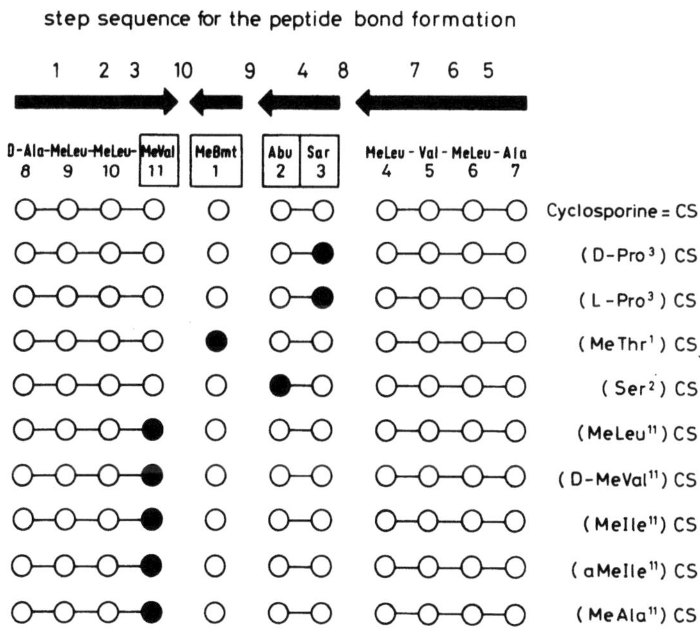

Fig. 2. Synthetic analogues of cyclosporine (strategy for their synthesis)

the cyclosporine molecule was expected to provide clues as to the type of interaction occurring between MeLeu-11 and the cyclosporine receptor on the lymphocyte. Other variations in position 11, structurally closer to cyclosporine itself, were made by replacing MeVal by MeIle (N-methyl-L-isoleucine) and aMeIle (N-methyl-L-alloisoleucine), two amino acids having one additional methyl group on their carbon chain. In MeIle a methyl group replaces a hydrogen atom of the pro-*S* methyl of MeVal and in aMeIle a methyl group replaces a hydrogen of the pro-*R* methyl of MeVal. A further variation was the introduction in position 11 of a less hindered and smaller amino acid, namely N-methyl-alanine, instead of MeVal thus replacing the two methyl groups of MeVal by two hydrogen atoms.

With the exception of (L-Pro[3])cyclosporine, all of the above modifications would be expected to preserve the peptide ring conformation of cyclosporine and therefore changes in th biological activity due to variation of the alkyl groups of specific amino

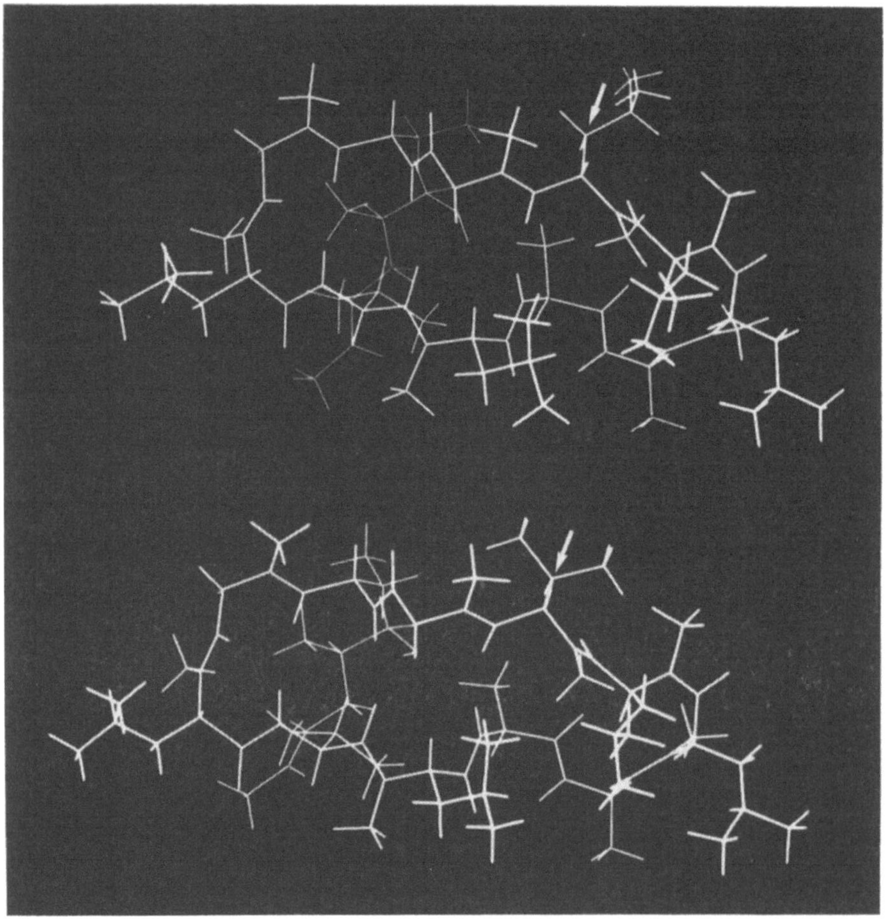

Fig. 3. A comparison of the 3 dimensional structures of cyclosporine (below) and synthetic (MeLeu[11])cyclosporine (above) as determined by X-ray crystallographic analysis (→ β-carbons of amino acids 11 indicated.)

acids should reflect whether or not these amino acids are involved in interactions with the receptor.

3.2 Synthesis of the Cyclosporine Analogues

The same synthetic strategy was used for the analogues as for the synthesis of cyclosporine itself (Scheme 5), the new peptide fragments being constructed as shown in Fig. 2. The numbers and arrows indicate the step sequence and direction of chain elongation. Black circles in the figure show the positions of the modified amino acids. Cyclisation was effected on the undecapeptides with a free amino group and free carboxy group using $(PrPO_2)_3$[44] in methylene chloride in the presence of 4-dimethylaminopyridine.

3.3 Structural Assignment of the Cyclosporine Analogues

The structural assignment for these cyclosporine analogues (Fig. 2) have been supported by nuclear magnetic resonance spectroscopy. All derivatives, except for (L-Pro^3)cyclosporine and (D-MeVal^{11})cyclosporine, the latter showing a complicated NMR-spectrum indicating at least 7 peptide ring conformations, gave NMR spectra similar to that of cyclosporine itself. They showed essentially only one conformation in deuterochloroform and only the expected minor changes in the chemical shifts for resonances of protons in the near environment of the structural change. The structural assignment for the (MeLeu^{11})cyclosporine has been further supported by an X-ray crystallographic analysis which confirmed the structure predicted by synthesis and spectroscopic measurements. It also showed that the peptide backbone of (MeLeu^{11})cyclosporine has practically the same conformation as cyclosporine itself in the solid state. The beta-carbons of the amino acids methylvaline-11 and methylleucine-11 are in the same relative position in the two molecules (Fig. 3).

3.4 Biological Activity and Structure-Activity Relationships

Figures 4 and 5 show the relative immunosuppressant activity of some synthetic and naturally occurring analogues of cyclosporine (*** indicates potent activity, ** intermediate activity, and little or no activity). The compounds were tested using pharmacological models previously used to test cyclosporine[2-6] included the suppression of antibody-forming cells in the spleens of mice immunized against sheep erythrocytes, the oxazolone-induced skin reaction in mice, and the Freund adjuvant arthritis model.

From the results presented in Fig. 4 it is clear that the carbon chain of amino acid 1 is important for biological activity. Removal of the non-polar part of the side chain as in (MeThr^1)cyclosporine reduces the immunosuppressive activity dramatically as does modification of the hydroxy group as seen in O-acetyl- and desoxy-cyclosporine. Even the double bond makes a contribution to the biological activity since

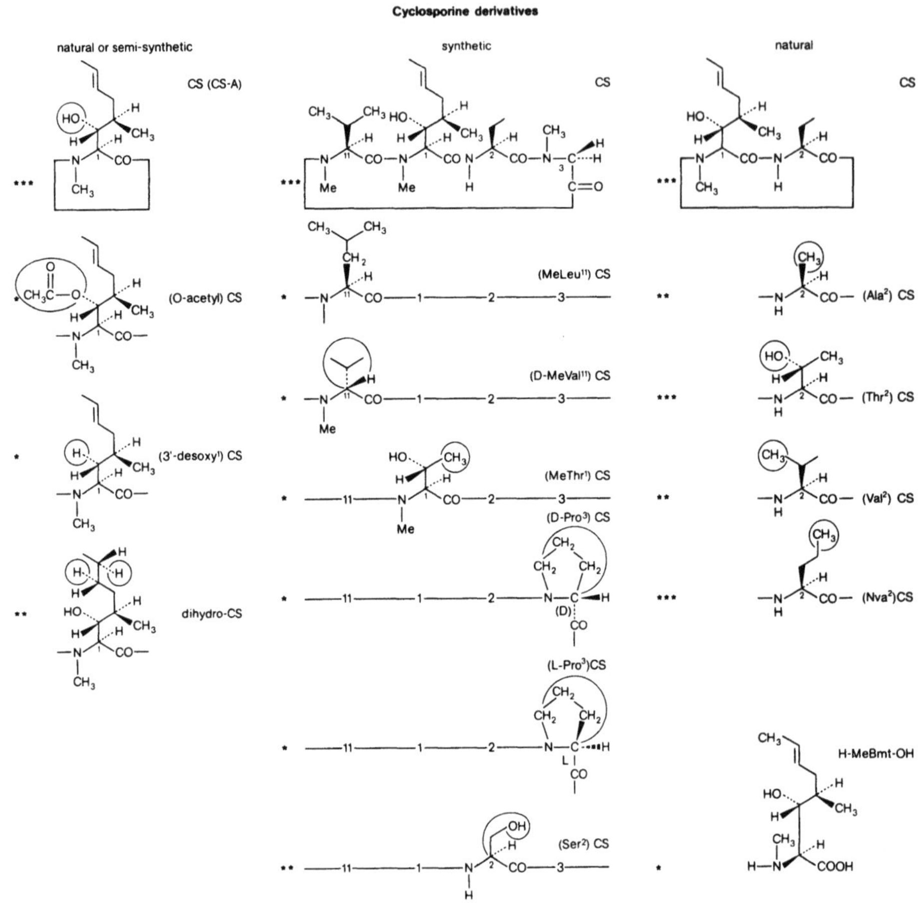

Fig. 4. Immunosuppressive activity of some cyclosporine derivatives. CS = cyclosporine; *** Strong immunosuppressive activity; ** intermediate activity; * little or no activity

dihydro compounds are generally less potent. The novel amino acid MeBmt alone is not sufficient for biological activity.

Amino acid 2 permits some variation with alkyl chains of 2–3 carbons showing good activity. The smaller chain in (Ala^2)cyclosporine and the more bulky residue in (Val^2)cyclosporine lead to a small reduction in potency. Although (Thr^2)cyclosporine is a potent immunosuppressant, (Ser^2)cyclosporine is significantly less active suggesting that the hydroxy group is not involved in a strong interaction with the receptor and that the hydrophobic interactions between the alkyl chain of amino acid 2 and the receptor are more important.

Although (D-Pro^3)cyclosporine and (MeLeu^{11})cyclosporine should have almost the same conformation as cyclosporine itself (similar NMR spectra for both derivatives and similar X-ray crystal structure of the latter with cyclosporine) they are negative variants suggesting that in this case the additional steric bulk of these peptide chains in position 3 and 11 respectively, probably prevents effective binding of these derivatives to the cyclosporine receptor. The lack of activity of the

(L-Pro^3)cyclosporine is not unexpected. This compound not only incorporates additional steric bulk but also has a different conformation from that of cyclosporine, probably with a beta-turn of type II at position 3. Likewise the loss of activity observed with (D-MeVal^{11})cyclosporine is due to the conformation of the peptide ring which is certainly very different (NMR, X-ray) from that of cyclosporine at a position which appears very sensitive to minor structural variations (see below). It is not possible to draw detailed structure-activity relationships from such derivatives in which there is a great change in the overall shape of the molecule.

From the results presented in Fig. 5 it is clear that minor variations of the structure of amino acid 11 can reduce the immunosuppressive activity of cyclosporine analogues. The pro-R methyl group of N-methylvaline-11, which is closer to the N-methyl group of MeBmt on the basis of the X-ray crystal structure of cyclosporine, is somewhat more sensitive to variation than the pro-*S* methyl group. The lower activity of the MeAla analogue could be due to the failure to fill a critical hydrophobic cavity on the receptor or to reduced conformational rigidity of the molecule.

3.5 Conclusions

Some structure-activity relationships for cyclosporine are now emerging from the study of specifically modified derivatives. However, in order to obtain a more precise image of the importance of all the structural features of this molecule and then to prepare derivatives with an improved profile of activity, much remains to be done. It is nevertheless already clear that the unusual MeBmt chain is intimately

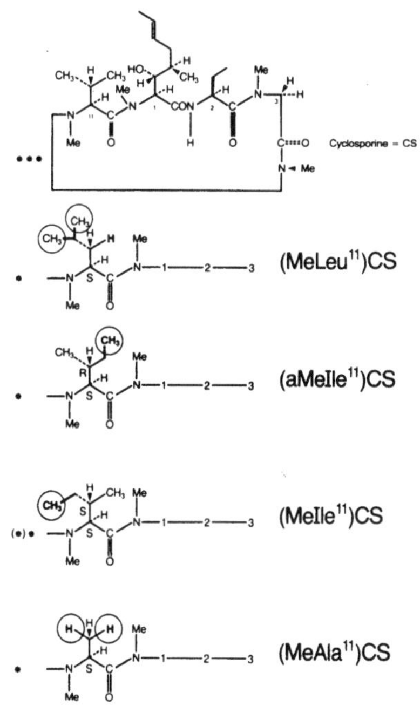

Fig. 5. Immunosuppressive activity of some synthetic cyclosporine analogues varied in position 11. *** Strong immunosuppressive activity; ** intermediate activity; * little or no activity

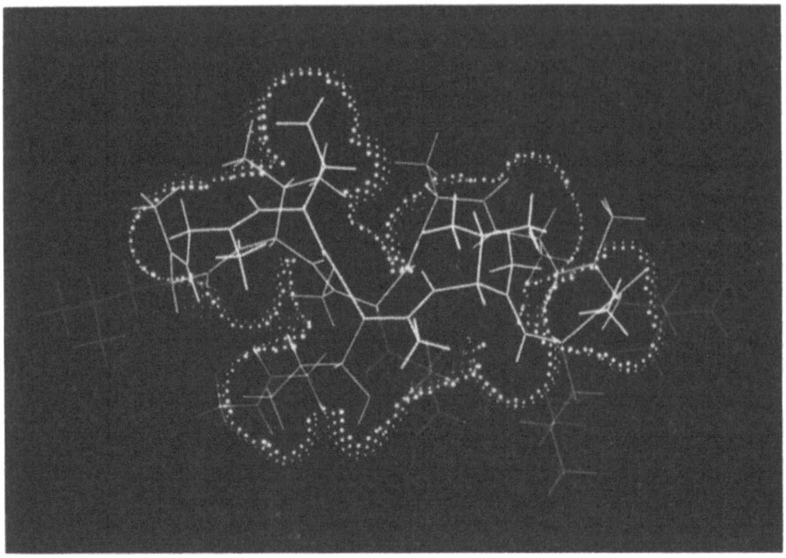

Fig. 6. Clip of a computer generated van der Waals surface of cyclosporine in the region of the amino acids varied synthetically showing at the same time the skeleton of amino acid 1 (MeBmt, below), 3 (Sar, left centre), 2 (Abu, above) and 11 (MeVal, right centre).

involved in the biological actions of the cyclosporine molecule but that this amino acid alone is not sufficient for immunosuppressive activity. The biological activity is associated with a larger portion of the cyclosporine structure and the results presented above indicate that this probably includes amino acids 1, 2, 3 and 11 (see Fig. 6).

4 The Mechanism of Action of Cyclosporine

4.1 Introduction

The basic observation that cyclosporine suppresses antibody formation and cell mediated immune responses was first reported in 1976[2]. Simple experimental systems indicated that this fungal metabolite acted reversibly and selectively on lymphocytes, mainly on T cells, and rather affected the induction phase than the proliferative phase of lymphoid cell populations.

These unusual properties of a fungal peptide triggered not only the extensive chemistry and studies on structure-activity relationships of cyclosporine as outlined in the previous sections, but also led to an extensive investigation on the mode of action of this immunosuppressive drug. The advances in understanding and dissecting the individual steps of the immune response allowed far more detailed studies to define the site of action of cyclosporine. Since the available experimental systems for the isolated study of individual processes in the cascade of events in lymphoid cell activation can still not be fully controlled, numerous inconsistent and in some cases conflicting results have accumulated in the extensive literature. Here we shall concentrate on those findings which we think are particularly relevant and appear well founded. The current working hypothesis which emerged from these findings will be compared with known and possibly analogous mechanisms as established for other agents and systems, which have been studied in more detail.

4.2 *Effect of Cyclosporine on Lymphoid Cells*

4.2.1 T Lymphocytes

Activation of T lymphocytes begins with the recognition of antigen in the context of class II histocompatibility antigens (e.g. I.A antigens of the mouse MHC) as expressed on antigen presenting cells such as macrophages or dendritic cells. This cell

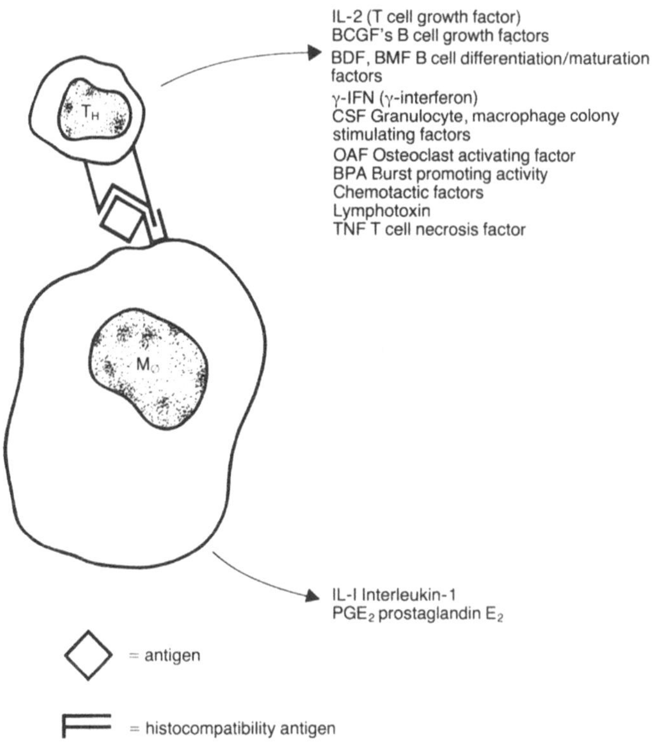

Fig. 7. T lymphocyte activation by cell interaction. Foreign antigen is recognised on the surface of an antigen presenting cell (M, macrophage) in association with the compatible histocompatibility antigen (class II for T helper cells, class I for cytolytic cells). The lymphokines produced affect the growth and maturation of a variety of cell types

interaction triggers a cascade of primary and secondary events on both cell types. Both macrophages and T cells express new or amplify the expression of existing surface structures and release soluble, biologically active mediators (lymphokines). The most significant new structure expressed on T lymphocytes is the receptor for the T cell growth factor (interleukin 2, IL-2). This IL-2 receptor expression which has been called stage 1 of T cell activation and occurs 6–10 h after activation, precedes the release of the growth factor IL-2. In this second stage, IL-2 production, the macrophage product interleukin 1 (IL-1) seems to be involved. T cell growth per se (stage 3) is then the result of the interaction of IL-2 and the IL-2 receptor, which has high affinity for its specific ligand ($K_m = 5–10 \times 10^{-11}$). IL-2 is only one of a large number of lymphokines which are released by T cells as a consequence of this activation process. Among them are macrophage activating factors, interferons, colony-stimulating factor, all of which exert positive effects on macrophages, like increasing expression of I.A antigens, cytolytic activity or growth. Other lymphokines affect growth of B lymphocytes and their maturation to antibody secretion. Since IL-2 acts on any T lymphocyte population which expresses the IL-2 receptor, interference with T cell activation or IL-2 production will also have serious consequences for IL-2 mediated T cell – T cell interactions. In view of the complexity of these early events in T cell activation it is not surprising that primary and secondary effects of cyclosporine are difficult to discern.

The early experiments of Borel et al.[4)] had shown that mitogen-induced T cell proliferation (Con A, PHA) could be abrogated when cyclosporine was added after 2 h but not after 48 h of T cell culture. The cells were blocked in (G_0 or) early G_1 phase of the cell cycle and these cells which had already progressed beyond the early G_1 phase still went on to complete their cycle in the presence of cyclosporine[46)].

Once the two key events in the T lymphocyte activation process, namely IL-2 receptor expression (stage 1) and IL-2 production (stage 2) could be discerned in newly established experimental systems, the effect of cyclosporine on the T cell response could be studied in greater detail. The first report on the topic by Larsson could not be confirmed. E.-L. Larsson's results had suggested that cyclosporine inhibited IL-2 receptor-expression but not the production of this lymphokine and therefore classified cyclosporine as a stage 1-inhibitor[47)]. All subsequent studies, including our own, showed clearly that cyclosporine shuts off IL-2 production[48–50)]. Such studies were carried out under a wide variety of experimental conditions using mitogenic or antigen specific stimulation of primary T lymphocyte cultures, T cell lines and antigen specific T cell clones of human and mouse origin. The effect on IL-2 receptor expression is still in part conflicting. The results of Palacios[48)] point to an inhibition of IL-2 receptor expression which could be due to reduced production of IL-1 once OKT4+ T lymphocytes (helper T cells) are blocked by cyclosporine. This model would suggest that IL-1 is involved in the expression of the IL-2 receptor. Functional studies by Bunjes et al.[49)], DosReis and Shevach[50)] and Cammisuli[59)] indicated that the expression of the receptor was unaffected while IL-2 production was abrogated in mouse and guinea pig lymphocytes.

Studies on the expression of IL-2 receptors with monoclonal antibodies, which bind selectively to the IL-2 receptor on human lymphocytes (anti-TAC-antibody) showed that IL-2 receptors were expressed in the presence of cyclosporine and that addition of IL-2 containing supernatants restored the proliferative responses of T lymphocytes (Miawaki et al.[51)]). By contrast, the results obtained by Lillehoj et al.[52)] using murine lymphocytes and the rat monoclonal antibody, 7D4, which recognizes a site on the interleukin 2 receptor distal to the interleukin 2 binding site, are more difficult to interpret. Cyclosporine inhibited the proliferative response and interleukin 2 production and although the cells expressed the IL-2 receptor after one day, a dose dependent inhibition of receptor expression by cyclosporine was seen after 3 days in culture following concanavalin A stimulation. More surprising was the fact that exogenously added IL-2 and/or IL-1 failed to restore the receptor expression and the proliferative response. These results suggest that cyclosporine possibly blocks the synthesis of more than one lymphokine and if it does inhibit interleukin 2 receptor expression, it does so by an indirect mechanism.

The inhibition of interleukin 2 synthesis by cyclosporine, which will be discussed at the sub-cellular level in more detail below, has obvious consequences in the sequence of events which follows T cell activation. As will be clear from the key role of this lymphokine (Fig. 7), proliferation of cytolytic T lymphocytes is abrogated in the presence of effective concentrations of cyclosporine[49, 53–55)], while their cytolytic effect is unaffected. This holds both for specific, class I MHC-antigen-restricted cytolytic T cells (killer-cells) which can produce their own IL-2 (helper-independent killer cells) and for cytolytic T cell precursors which clonally expand in response to IL-2 provided by helper T cells. Their generation in vitro in the presence of cyclosporine can be restored by exogenous IL-2.

The effect of cyclosporine on suppressor T lymphocytes is at present not fully understood. Suppression remains a difficult parameter to measure qualitatively and quantitatively in experimental in vitro systems. This may account for some of the discrepencies reported in the literature[56–64)]. In vivo experimental models strongly suggest that cyclosporine selectively spares or even favours the activation of T suppressor cells. Their soluble mediators have in fact been implied in allograft survival and abrogation of host effector responses against organ allografts. Since this topic has been comprehensively reviewed recently[64)], further discussion of the effects of cyclosporine on suppression appears of little value.

4.2.2 B Lymphocytes

The initial findings of Borel et al.[4)] indicated a strongly reduced antibody formation in cyclosporine treated mice as determined by serum haemagglutination titres and the appearance of specific antibody producing cells in the spleen. Also in guinea pigs a single dose of cyclosporine can suppress the production of antibodies to hapten and carrier in primary and secondary responses[56)]. These profound effects on T cell dependent B cell responses can be fully accounted by a selective effect of cyclosporine on helper T cells since it was shown that polyclonal B cell activation by the B cell mitogen lipopolysaccharide (LPS) was not affected at T cell suppressive concentrations of this drug[65)].

Studies by Klaus and his collaborators indicated that a subpopulation of B cells which respond to non-mitogenic Type 2 T-independent (TI-2) antigens is highly sensitive to inhibition by cyclosporine[66)]. It should however be noted that B cell responses to TI-2 antigens also require T cell derived growth and maturation factors.

Therefore the assumption of a cyclosporine sensitive B cell subpopulation has to exclude carefully an indirect mode of action of cyclosporine on virgin helper T cells. It also remains to be shown whether the B cell response to the Type I T-independent antigens (like DNP-LPS), which was found to be unaffected by cyclosporine, was truly induced by the hapten DNP or resulted from polyclonal B cell activation. The B cell response induced by anti-Ig-antibodies was also shown to be highly susceptible to inhibition by cyclosporine if the drug was added during the first 4 h. Further studies[66)] indicated that the anti-immunoglobulin induces B lymphocytes to enter the G_1 phase of their cycle and although cyclosporine could block this initial priming step, once the cells had entered the G_1 phase, they could no longer be inhibited. A similar priming of some B cells can be achieved with phorbol myristate acetate but this is resistant to cyclosporine. Using a crude B cell growth factor and anti-μ antibodies as stimulants in human B cells, cyclosporine was shown to inhibit the initiation of RNA synthesis. Taken together the above results suggest that it is only the priming step of B cell activation which is susceptible to inhibition by cyclosporine.

4.2.3 Monocytes, Granulocytes

Cyclosporine has no direct effect on granulocyte or macrophage function. However, synthesis and release of interleukin 1 (IL-1) by macrophages is a result of cell activation which is either mediated by cellular interaction with T lymphocytes or mediated by lymphokines such as colony stimulating factor, gamma-interferon, or other macrophage activating factors. Reduced production of IL-1 may therefore well be due to a cyclosporine effect on T cells[53, 67)]. Lipopolysaccharide induced release of IL-1 from macrophages is not inhibited by cyclosporine[53)].

4.3 General Mechanisms of Cell Activation

Normal lymphocytes require a combination of at least two external signals for activation. In T lymphocytes one signal is provided by the interaction of antigen and I.A molecules on the macrophage surface with the T cell receptor. Interleukin 1 is regarded as a second signal while the proliferation results from the interaction of interleukin 2 with its receptor. The mechanism by which these interactions with the surface membrane are translated into observed cellular responses is not fully understood. In some cell systems, interleukin 1 can be replaced by phorbol myristate acetate[68)] suggesting that the two agents may possess signal-response coupling mechanisms with some features in common.

4.3.1 Cyclic Nucleotides

The translation of activation signals on the exterior membrane surface into intracellular responses has been investigated widely in non-lymphoid systems in the past few years. Extensive studies have been published on the mechanism by which ligands acting on receptors coupled to adenylate cyclase can modulate intracellular cyclic AMP levels and thereby the level of phosphorylation and activity of specific proteins via the cyclic AMP dependent protein kinase. These processes are however not believed to be critical in the normal activation mechanism of T lymphocytes although the inhibitory actions of some prostaglandins[69] could involve cyclic AMP dependent processes. Cyclosporine does not signiflcantly alter cyclic AMP or cyclic GMP levels in lymphocytes[70] and its mode of action clearly involves other processes.

4.3.2 Receptor Activated Calcium Influx

Receptor activated calcium gating is certainly an important mechanism in the activation of lymphocytes. Increases in the intracellular calcium concentration can result from membrane depolarisation followed by calcium influx through potential sensitive channels in lymphocytes[71]. However, powerful inhibitors of potential dependent channels (e.g. the dihydropyridines such as nifedipine) generally fail to inhibit lymphocyte activation except at high concentrations. It is therefore unlikely that intracellular calcium concentrations are raised in lymphocytes by this mechanism.

Studies with ligands activating receptor operated channels indicate that the interaction with the receptor leads to the activation of a phosphatidylinositol-4,5-diphosphate specific phospholipase C which hydrolyses its substrate to inositol-1,4,5-triphosphate and diacyl glycerol[72]. The phospholipase C is rate limiting as kinases rapidly replentish supplies of substrate by phosphorylating phosphatidylinositol to phosphatidylinositol-4-phosphate and then to phosphatidylinositol-4,5-diphosphate thereby restoring the normal equilibrium between these lipids (see Fig. 8). Both products of hydrolysis have biological functions. Inositol-1,4,5-triphosphate releases calcium which is bound to internal stores thereby raising intracellular calcium ion concentrations and activating calcium and calmodulin dependent kinases and other proteins. Diacylglycerol activates a calcium and phospholipid-dependent kinase, protein kinase C, before it itself is deactivated by phosphorylation to phosphatidic acid. The diacyl glycerol binding site on protein kinase C has also been established as the receptor for the tumor promoting phorbol esters such as phorbol myristate acetate[73], the latter compound acting as a potent and long acting diacylglycerol agonist. Lymphocytes contain appreciable quantities of protein kinase C[74] and although the role of this kinase in the activation processes has not yet been defined, it may well be related to the mode of action of interleukin 1 which has a similar broad spectrum of activities. Diacylglycerol (or phorbol myristate acetate) and calcium ions act synergistically in many experimental systems of cell activation[75]. Both in T and B lymphocytes the combined effect of calcium ionophores with phorbol esters has now been demonstrated to be a sufficient activation signal[76, 77]. Receptor activated calcium influx may involve additional mechanisms apart from the potential dependent channel and inositol-1,4,5-triphosphate mediated processes. Where exchange of calcium in both directions across the membrane occurs for long periods, the average intracellular calcium ion concentration, as measured by the quin-2 method, may not reflect the true situation as calcium ion concentrations are probably only high in the immediate vicinity of the inner surface of the membrane.

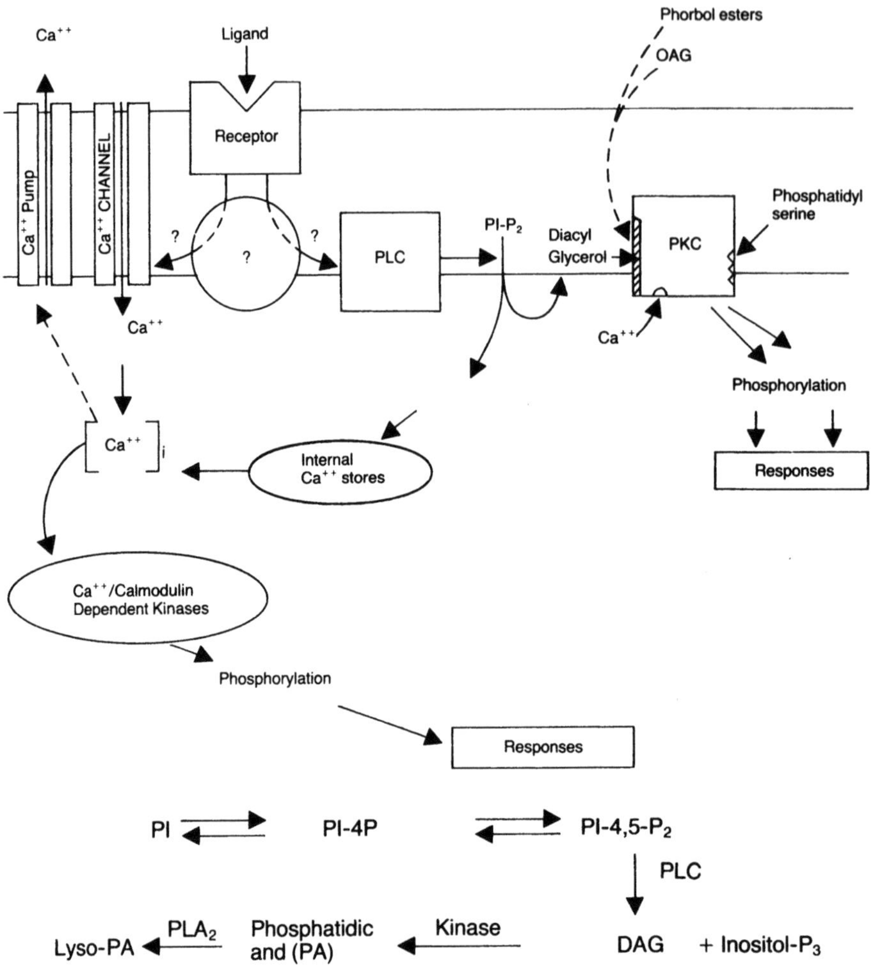

Fig. 8. Phosphatidyl inositol turnover and receptor operated channels. Ligand-receptor binding leads to the release of Ca^{++} from internal stores via a pathway involving hydrolysis of phosphatidylinositol-4,5-diphosphate (PI-4,5-P_2) by phospholipase C (PLC) to diacylglycerol (DAG) and inositol-1,4,5-triphosphate (Inositol-P_3). A further increase in the cytosolic free Ca^{++} concentrations ($[Ca^{++}]_i$) occurs via influx. Phosphatidylinositol-4,5-diphosphate is in equilibrium with phosphatidylinositol-4-phosphate (PI-4P) and phosphatidylinositol (PI). Phorbol esters and 1-oleoyl-2-acetylglycerol can by-pass the receptor and activate protein kinase C (PKC) directly

Cyclosporine does not inhibit calcium influx into lymphocytes induced by concanavalin A[78], indicating that the action of the drug is not at this level.

4.3.3 Growth Factors and Hormones

Another recently discovered signal response coupling mechanism involves activation of tyrosine specific protein kinases. The receptors for epidermal growth factor, insulin and platelet derived growth factor have tyrosine kinase activity as a functional unit incorporated into their receptor protein[79–81]. Upon binding their ligands this type of receptor can phosphorylate specific membrane proteins on tyrosine. They can also phosphorylate themselves or have their activity modulated via phosphorylation on serine by other kinases like protein kinase C and become internalized permitting them to phosphorylate intracellular proteins. The epidermal growth factor receptor possesses in addition endonuclease activity[82] which provides a link to a nuclear site of action and a direct involvement of such factors on transcriptional control.

It has been shown that lymphocyte stimulation leads to phosphorylation of proteins on tyrosine and the addition of a large excess of a synthetic, low molecular weight peptide substrate of a tyrosine kinase inhibits T cell activation by concanavalin A[83]. One of the targets of tyrosine kinases in activated lymphocytes is lipomodulin, the endogenous phospholipase A_2 inhibitory protein, which is deactivated in the phosphorylated state[84]. The influence of cyclosporine on tyrosine kinase activity in lymphocytes has not yet been studied.

Several hormones have been shown to act intracellularly and to induce the synthesis of a few specific proteins. Prolactin binds to receptors on the membrane surface and is then endocytosed and degraded. It is believed that a small fragment then acts as an intracellular messenger[85]. A similar mechanism has been proposed for insulin in addition to the tyrosine kinase concept given above.

The corticosteroids (and other steroids) represent another example of hormone-induced synthesis of a small number of defined proteins. The steroid appears to penetrate the plasma membrane, presumably binding to a cytosolic receptor at the membrane inner surface. The receptor-steroid complex is then phosphorylated and transported to the nucleus where it binds to a specific promoter sequence of DNA inducing mRNA-transcription and translation for a number of proteins[86–89]. The mode of action of steroids is still controversial in some aspects[90]. The mechanisms by which hormones and growth factors modulate the synthesis of specific proteins may be related to the mechanism of action of cyclosporine as discussed below.

4.4 Membrane and/or Cellular Receptors

Binding studies with tritiated dihydrocyclosporine C have revealed cyclosporine receptors on the surface of mouse and human lymphocytes[91, 92]. These binding sites have a K_D of approximately 10^{-7} M, which corresponds closely to the concentrations of CsA effective in lymphocyte cultures. Similar cyclosporine binding sites also exist on other cell types such as polymorphonuclear leukocytes and macrophages, which are not affected by cyclosporine. The expression of binding sites for a ligand, which has no apparent effect on cell function is not uncommon, since the presence of beta-receptors on turkey erythrocytes or human lymphocytes has been shown.

The observation that cyclosporine bound in a similar saturable and reversible manner to phospholipid vesicles (prepared from phosphatidylcholine and cholesterol in a 1:1 ratio) as it did to lymphocytes, raised more serious doubts as to the relevance of these binding studies. On the basis of these results LeGrue et al.[93] argued against the existence of a specific surface receptor and concluded that cyclosporine simply partitioned into the membrane and thereby altered membrane fluidity. In their view, differences in binding of various cyclosporine derivatives to lymphocytes and phospholipid vesicules reflected their differences in partition coefficients. A possible criticism of the experiments with the phospholipid vesicules is the high proportion of cholesterol used which might be expected to promote the partitioning of a lipophilic drug into the membrane. Also, it is difficult to reconcile a mechanism of action based on a general property like altered membrane fluidity with the structure-activity relationships as shown in the previous section: minor structural variations on the cyclosporine molecule which hardly affect lipid solubility but have a dramatic influence on immunosuppressive action. A certain correlation between the ability of cyclosporine derivatives to displace lymphocyte bound CsA and their immunosupressive potency, as observed by LeGrue, suggests membrane penetration may be one, but not the only requirement for biological activity.

Studies on the turnover of lymphocyte membrane lipids by Resch and his collaborators[94)] may yet provide another explanation for cyclosporine induced changes in membrane properties which could be relevant to its mode of action. During the activation process lymphocytes take up unsaturated fatty acids from the medium and incorporate them into their membrane phospholipids. A detailed study has revealed that an acyl residue is removed from the 2 position of phosphatidylcholine by a phospholipase A_2 and replaced by an unsaturated fatty acid residue. In the absence of a source of unsaturated fatty acids, growth and cell proliferation cannot occur[95)]. Resch also showed that cyclosporine inhibited the lipid turnover in lymphocytes at concentrations slightly above those required for immunosuppression and postulated that the mechanism of action of cyclosporine could involve inhibition of the acyl transferase[94)]. Although this enzyme is common to most cells and its direct inhibition by cyclosporine would not explain the cellular specificity of this drug, the results deserve some attention as changes in phospholipid composition could alter the properties of some surface receptors. It could explain the effects of cyclosporine on IL-1 and antibodies to lymphocyte surface structures described below. Similarly, the phospholipase A_2 inhibitory protein lipomodulin, which is induced by the action of corticosteroids, could presumably inhibit lymphocyte activation by preventing the phospholipid breakdown and as a consequence the incorporation of unsaturated fatty acids into the membrane. Very high concentrations of cyclosporine inhibit pancreatic phospholipase A_2 in an assay using a synthetic substrate[96)]. However, these effects cannot be observed with membrane substrates and phospholipase A_2 or on arachidonic acid release from intact cells other than macrophages[96)]. The findings of Resch suggest that cyclosporine inhibits the mechanism which activates the phospholipid turnover. Phospholipase A_2 activity is normally regulated by lipomodulin phosphorylation and/or calcium influx. Since the latter is not affected by cyclosporine, a critical kinase activity may be directly or indirectly inhibited by the drug.

Studies on the binding of monoclonal antibodies to surface structures of human T cells revealed that cyclosporine inhibited the binding of the monoclonal antibody directed agaist the surface structure T3 while the binding to T4 and T8 was unaffected[97)]. T3 is a 20/25 kD glycoprotein which is present on all mature human T cells and is closely associated with the specific T cell recognition structure, a 90 kD hetero-dimer termed Ti[98)]. The interaction of this T3.Ti molecular complex with antigen in the context of Class II histocompatibility antigenes activates T cells through the IL-2 dependent autocrine pathway outlined in Sect. 4.2.1. Further studies by Ryffel indicated that preincubation of cells for one hour with concanavalin A, phytohaemagglutinin and OKT3 antibodies significantly inhibited the subsequent binding of dihydro-cyclosporin C, whereas the specifically bound cyclosporine derivative was not displaced by these and other mitogens or interleukin 2[92]. Both Palacios and Ryffel concluded from these observations that cyclosporine binds to or near the receptor which is involved in T cell activation. In another series of experiments, Palacios showed that T cells became unresponsive to interleukin 1 after treatment with cyclosporine. More recently, Bendtzen and Dinarello[99] showed that cyclosporine, but not an inactive analogue, antagonised the apparent absorption of highly purified interleukin 1 to T lymphocytes and not to B cells or erythrocytes. Although specific receptors for interleukin 1 have not been demonstated on any cell type, the observed absorption and displacement was selective and reversible.

A cytosolic receptor protein, designated cyclophilin[100], has recently been shown to be responsible for the concentration of CsA by lymphoid cells. Purified to homogeneity from bovine lymphocytes, cyclophilin binds cyclosporine in a 1:1 ratio with a dissociation constant of approximately 2×10^{-7} M. A partial amino acid sequence of this 15 kD protein has no apparent homology with known protein sequences. The affinity of a series of cyclosporine derivatives to cyclophilin correlates with their relative immunosuppressive activity[100]. The two types of binding studies with cyclosporine derivatives may suggest that the lipophilic agent traverses the membrane and then is captured and transported by a carrier protein through the cytosol of the lymphocyte and possibly to the nucleus. The quantity of cyclosporine actually reaching the nucleus may however be very small since the bulk of fluorescence labelled cyclosporine derivates accumulate in lipid vesicules[101].

Another hypothesis on the mode of action of cyclosporine is based on the findings of Larson and Russell[102] which showed that cyclosporine can displace prolactin from its receptor present on lymphocytes. Cyclosporine also blocked the prolactin induced ornithine decarboxylase activity in lymphocytes and kidney cells. The role of prolactin or a prolactin like factor such as proliferin[103] in lymphocyte activation is currently under investigation. It should however be noted that displacement of prolactin from its binding site on lymphocytes by cyclosporine does not necessarily imply that the two compounds compete for the same receptor. Indirect mechanisms, such as induction of receptor phosphorylation, can lead to ligand displacement as is the case for the displacement of epidermal growth factor by phorbol esters[104].

4.5 Cyclosporine and Transcriptional Control

Immunosuppression induced by cyclosporine results at least in part from the inhibition of lymphokine secretion by T cells which provide the necessary growth and differentiation signals for B cells, T cells and macrophages[49, 105–107]. One of these factors is interleukin 2, the principle stimulus in causing the proliferation of activated T lymphocytes. Studies on IL-2 gene expression with the cloned leukemic T cell line Jurkat and a cDNA probe for human IL-2 have shown[108] that cyclosporine inhibits the expression of mRNA for IL-2 but not of two other genes, the IL-2 receptor and HT-3, which are also inducible on stimulation with phytohemagglutinin (PHA) and phorbol myristate acetate (PMA). Preincubation of the cells with cyclosporine or concommitant administration together with the stimulus almost completely inhibited IL-2 mRNA production. Addition of cyclosporine only 4 h after the PHA and PMA had no effect although IL-2 transcription occurs for at least 15 h following induction[108]. The isolated nuclei from cyclosporine treated cells also showed reduced IL-2 transcription whereas the constitutive HLA transcriptional activity was only slightly inhibited. These observations suggest a nuclear site of action whereby cyclosporine selectively blocks a process leading to the triggering of transcription of a small number of gene products. Similar results were obtained by Elliott et al.[109] who translated the mRNA from cyclosporine treated und untreated phorbol myristate acetate induced EL-4 cells in oocytes and quantitated the translated interleukin 2 in the biological assay. These studies showed also that the induction of less than one percent of mRNA's expressed by these cells are sensitive to inhibition by cyclosporine and that only a few (6 to 8) induced proteins fail to show up in two dimensional gel-electrophoresis of cellular proteins.

Immunosuppressive concentrations of cyclosporine have also been shown to inhibit RNA polymerase II in a membrane free system[110]. In order to gauge the relevance of this finding it will be necessary to test a series of active and inactive analogues. In addition, an explanation for the selective action on polymerase II of lymphocytes, such as a specific transport mechanism, will have to be found.

4.6 Conclusions and Outlook

In our present understanding of the mode of action of cyclosporine the abrogation of the production of IL-2 and other lymphokines appears to account for the majority of its immunosuppressive effects. Binding studies and the nature of the receptor are still controversial but point to a transport through membrane and cytosol to nuclear control elements of transcription.

Questions can now be asked as to which other genes, besides IL-2, are prevented from transcription and to define their role in lymphocyte function? These questions are likely to be answered in the near future but it seems reasonable to assume that the other genes which fail to be transcribed in the presence of cyclosporine code for other lymphokines. These include osteoclast activating factor[111], gamma-interferon[112], and monocyte chemotactic factor[113], colony stimulating factor[114], lymphotoxin, T cell necrosis factor, and chemotactic factors. When the nature of all these lymphokines is known, it will be then possible to determine which actions of cyclosporine are direct and which are consequences of abrogated lymphokine production. This should help to define the primary molecular target of cyclosporine between surface receptors on the lymphocyte and transcriptional control elements on DNA. It is tempting to compare the mechanism of action of CsA to that proposed for the corticosteroids, where binding to a receptor protein, transport through the cytosol and subsequent binding to and activation of a DNA sequence by the steroid-receptor complex appears to be well established[89] (c.f. Fig. 9). This corticosteroid-induced chain of events lead to the induction of some specific proteins whereas in the case of cyclosporine, the action of the putative cyclosporine-cyclophilin complex may repress some gene sequences which are normally activated during T-lymphocyte stimulation. Peptide growth factors such as epidermal growth factor, platelet derived growth factor and insulin are known to selectively induce[115] or inhibit[116] the mRNA synthesis for a number of proteins.

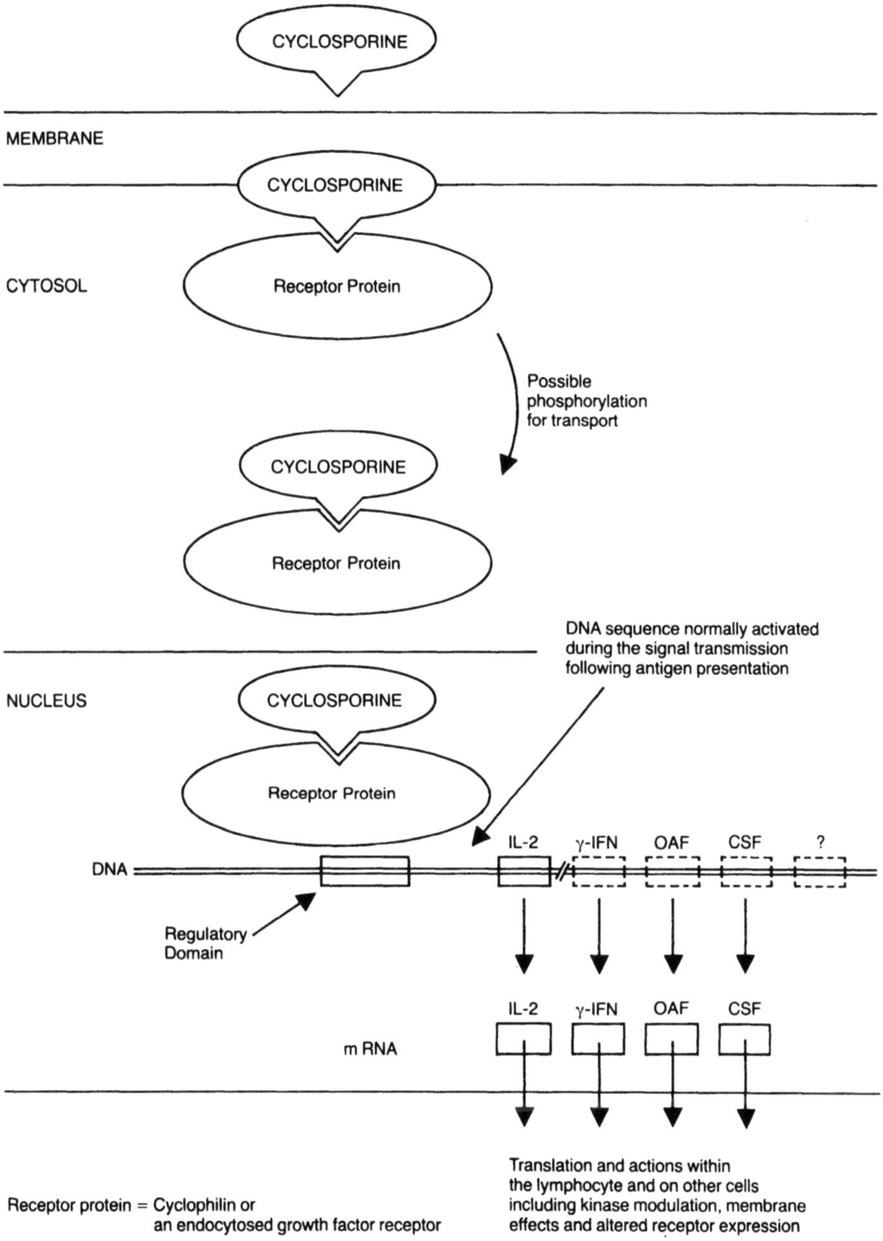

Fig. 9. A hypothetical model for the mode of action of cyclosporine. Receptor protein = cyclophilin or an endocytosed growth factor receptor

Cyclosporine may act as an agonist or antagonist of a peptide growth factor. The very specific structure-activity relationships observed with cyclosporine derivates would suggest a unique binding site for the drug, and therefore an endogenous regulatory molecule acting at this site. Otherwise nature would hardly have conserved the cyclosporin binding protein over a wide variety of animal species. The isolation of sufficient cyclophilin or the production of monoclonal antibodies to the active and inactive face of cyclosporine may provide the means of isolating such a regulatory factor.

5 References

1. For practical reasons it was previously proposed to define "cyclosporin" as the cyclic undecapeptide with the structure of "cyclosporin A" (Wenger, R. M.: Chemistry of Cyclosporin, in: *Cyclosporin A,* (ed.) White, D. J. G., p. 19, Elsevier Biomedical Press, Amsterdam (1982); Wenger, R. M.: *Helv. Chim. Acta 67,* 502 (1984)). Now that the USAN name "cyclosporine" has been accepted in the USA, this name will be adopted for the basic structure
2. Borel, J. F., Feurer, C., Gubler, H. U., Stähelin, H.: *Agents Actions 6,* 468 (1976)
3. Borel, J. F., Feurer, C., Magnée, C.: *Immunology 31,* 631 (1976)
4. Borel, J. F., Feurer, C., Magnée, C., Stähelin, H.: *Immunology 32,* 1017 (1977)
5. Borel, J. F., Wiesinger, D., Gubler, H. U.: *Europ. J. Rheumatol. Inflammation 1,* 97 (1978)
6. Wiesinger, D., Borel, J. F.: *Immunobiol. 156,* 454 (1979)
7. Bueding, B. E., Hawkins, J., Cha, Y. N.: *Agents Actions 11,* 380 (1981)
8. Thommen-Scott, K.: *Agents Actions 11,* 770 (1981)
9. Morris, P. J.: *Transplantation 32,* 349 (1981)
10. Morris, P. J.: *Advances in Surgery 17,* 99 (1984)
11. Calne, R. J., White, D. J. G., Thiru, S., Evans, D. B., McMaster, P., Dunn, D. C., Craddock, G. N., Pentlow, D. B., Rolles, K.: *Lancet ii,* 1323 (1978)
12. Powles, R. L., Barrett, A. J., Clink, H., Kay, H. E. M., Sloane, J., McElwain, T. J.: *Lancet ii,* 1327 (1978)
13. Stiller, C. R. and Keown, P. A.: Cyclosporin therapy in perspective, in *Progress in Transplantation,* P. J. Morris and N. L. Tilney, ed., *1,* 11 (1984)
14. Ruegger, A., Kuhn, M., Lichti, H., Loosli, H. R., Huguenin, R., Quiquerez, C., von Wartburg, A.: *Helv. Chim. Acta 59,* 1072 (1976)
15. Dreyfuss, M., Harri, E., Hofmann, H., Kobel, H., Pache, W., Tscherter, H.: *Eur. J. Appl. Microbiol. 3,* 125 (1976). The producing organism which was originally identified as Trichoderma polysporum is now classified as Tolypocladium inflatum Gams (see ref. 19)
16. Traber, R., Kuhn, M., Ruegger, A., Lichti, H., Loosli, H. R., von Wartburg, A.: *Helv. Chim. Acta 60,* 1247 (1977); Traber, R., Kuhn, M., Loosli, H. R., Pache, W., von Wartburg, A.: *Helv. Chim. Acta 60,* 1568 (1977)
17. Traber, R., Loosli, H. R., Hofmann, H., Kuhn, M., von Wartburg, A.: *Helv. Chim. Acta 65,* 1655 (1982)
18. Traber, R., Hofmann, H., Loosli, H. R., von Wartburg, A.: 24th Interscience Conference on Antimicrobial Agents and Chemotherapy, Washington D.C., 8–10 Oct. 1984, Poster presentation, Abstract No. 900; *Helv. Chim. Acta* (manuscript in preparation)
19. Kobel, H., Traber, R.: *Europ. J. Appl. Microbiol. Biotechnol. 14,* 237 (1982)
20. Petcher, T. J., Weber, H. P., Ruegger, A.: *Helv. Chim. Acta 59,* 1480 (1976)
21. Loosli, H. R., Kessler, H., Oschkinat, H., Weber, H. P., Petcher, T. J., Widmer, A.: *Helv. Chim. Acta 68,* 682 (1985)
22. The new abbreviation is, in accordance with conventional practice, to be understood for an amino acid of the L-configuration. The Me represents the N-methylated residue of Bmt in the same way as MeLeu is used for N-methyl-L-leucine. Bmt is the abbreviation of (4*R*-4-((*E*)-2-butenyl)-4-methyl-L-threonine

23. Chou, P., Fasman, G. D.: *J. Mol. Biol. 115,* 135 (1977)
24. Smith, J. A., Pease, L. G.: *CRC Critical Reviews in Biochemistry 8,* 316 (1980)
25. Richardson, J. S.: *Advances in Protein Chemistry 34,* 167 (1981)
26. Kobel, H., Loosli, H. R., Voges, R.: *Experientia 39,* 873 (1983)
27. Zocher, R., Kleinkauf, H.: *Biochem. Biophys. Res. Commun. 81,* 1162 (1978)
28. Zocher, R., Madry, N., Peeters, H., Kleinkauf, H.: *Phytochemistry 23,* 549 (1984)
29. Wood, A. J., Maurer, G., Niederberger, W., Beveridge, T.: *Transplant. Proc. 15* (suppl. 1), 2409 (1983)
30. Maurer, G., Loosli, H. R., Schreier, E., Keller, B.: *Drug Metabolism and Disposition 12,* 120 (1984)
31. Wenger, R. M.: *Helv. Chim. Acta 66,* 2308 (1983)
32. Cahn, R. S., Ingold, Ch., Prelog, V.: *Angew. Chem. 78,* 413 (1966); ibid *Int. Ed. 5,* 385 (1966)
33. Seebach, D., Hungerbühler, E.: Synthesis of enantiomerically pure compounds. Tartaric acid, an ideal source of chiral building blocks for synthesis, in: *Modern Synthetic Methods,* (ed.), Scheffold, R., Salle & Sauerlander, Frankfurt-Aarau, p. 93 (1980)
34. Seeley, D. A., McElwee, H.: *J. Org. Chem. 38,* 1691 (1973)
35. Herr, R. W., Wieland, D. M., Johnson, C. R.: *J. Am. Chem. Soc. 92,* 3813 (1970)
36. Schlosser, M., Christmann, K. F.: *Ann. Chem. 708,* 1 (1967)
37. Pfitzner, K. E., Moffatt, J. G.: *J. Am. Chem. Soc. 87,* 5670 (1965)
38. Wenger, R. M.: *Helv. Chim. Acta 66,* 2672 (1983)
39. Wenger, R. M.: *Helv. Chim. Acta 67,* 502 (1984)
40. McDermott, J. R., Benoiton, N. L.: *Can. J. Chem. 51,* 2551 and 2562 (1973)
41. Zaoral, M.: *Collect. Czech. Chem. Commun. 27,* 1273 (1962)
42. König, W., Geiger, R.: *Chem. Ber. 103,* 788 (1970)
43. Castro, B., Dormoy, J. R., Evin, J. G., Selve, C.: *Tetrahedron Lett. 14,* 1219 (1975)
44. Wissmann, H., Kleiner, H. J.: *Angew. Chem. 92,* 129 (1980): ibid *Int. Ed.* p. 133 (1980)
45. Kovacs, J.: in *The Peptides,* Vol. 2 (ed.), Gross, E., Meienhofer, J., Academic Press, New York–London, 1980, p. 485.
46. Koponen, M., Grieder, A., Loor, F.: *Exp. Cell Res. 140,* 237 (1982)
47. Larsson, E.-L.: *J. Immunol. 124,* 2828 (1980)
48. Palacios, R.: *J. Immunol. 128,* 337 (1982)
49. Bunjes, D., Hardt, C., Rollinghoff, M., Wagner, H.: *Eur. J. Immunol. 11,* 657 (1981)
50. DosReis, G. A., Shevach, E. M.: *J. Immunol. 129,* 2360 (1982)
51. Miyawaki, T., Yachic, A., Ohzeki, S., Nagaoki, T., Taniguchi, N.: *J. Immunol. 130,* 2737 (1983)
52. Lillehoj, H. S., Malek, T. R., Shevach, E. M.: *J. Immunol. 133,* 244 (1984)
53. Bunjes, D., Hardt, C., Solbach, W., Deusch, K., Röllinghoff, M., Wagner, H.: Studies on the mechanism of action of cyclosporin A in the murine and human T-cell response in vitro, in: *Cyclosporin A,* (ed.), White, D. J. G., p. 261, Elsevier Biomedical Press, Amsterdam (1982)
54. Andrus, L., Lafferty, K. J.: *Scand. J. Immunol. 15,* 449 (1982)
55. Morris, P. J., Mason, D. W., Hutchinson, I. V.: *Transplant. Proc. 15,* 2287 (1983)
56. Parker, D., Drossler, K., Turk, J. L.: *Int. J. Immunopharmac. 6,* 67 (1984)
57. Thomson, A. W., Moon, D. K., Nelson, D. S.: *Clin. Exp. Immunol. 52,* 599 (1983)
58. Thomson, A. W., Moon, D. K., Ingue, Y., Geczy, C., Nelson, D. S.: *Immunol. 48,* 301 (1983)
59. Cammisuli, S.: The effect of cyclosporin A on cell interactions within the immune system, in: *Cyclosporin A* (ed.) White, D. J. G., p. 243, Elsevier Biomedical Press, Amsterdam (1982)
60. Paarvonen, T., Järveläinen, H., Kontiainen, S., Häyry, P.: *Clin. Exp. Immunol. 43,* 342 (1981)
61. Palacios, R.: *Immunobiology 160,* 321 (1981)
62. Leapman, S. B., Filo, R. S., Smith, E. J., Smith, P. G.: *Transplant. Proc. 13,* 405 (1981)
63. Hess, A. D., Tutschka, P. J., Santos, G. W.: The effect of cyclosporin A on T-lymphocyte subpopulations, in: *Cyclosporin A* (ed.) White, D. J. G., p. 209, Elsevier Biomedical Press, Amsterdam (1982)
64. Kunkl, A., Klaus, G. G. B.: *J. Immunol. 125,* 2526 (1980)
65. Dongworth, D. W., Klaus, G. G. B.: *Eur. J. Immunol. 12,* 1018 (1982)
66. Klaus, G. G. B., Hawrylowicz, C. M.: *Eur. J. Immunol. 14,* 250 (1984)
67. Palacios, R.: *J. Immunol. 128,* 337 (1982)

68. Rosenstreich, D., Mizel, S. B.: *J. Immunol. 123,* 1749 (1979)
69. Bray, M. A.: *Immunol. Today 1,* 65 (1980)
70. Reinert, H., Loor, F.: unpublished results
71. Taylor, M. V., Metcalfe, J. C., Hesketh, T. R., Smith, G. A., Moore, J. P.: *Nature 312,* 462 (1984)
72. Berridge, M. J.: *Biochem. J. 220,* 345 (1984)
73. Kikkawa, U., Takai, Y., Tanaka, Y., Miyake, R. Nishizuka, Y.: *J. Biol. Chem. 258,* 11442 (1983)
74. Nishizuka, Y., Takai, Y., in: *Protein Phosphorylation,* Book A, (ed.) Rosen, O. M., Krebs, E. G., p. 237, Cold Spring Harbor Conferences on Cell Proliferation, Vol. *8* (1982)
75. Yamanishi, J., Takai, Y., Kaibuchi, K., Sano, K., Castagna, M., Nishizuka, Y.: *Biochem. Biophys. Res. Commun. 112,* 778 (1983); O'Flaherty, J. T., Schmitt, J. D., McCall, C. E., Wykle, R. L.: *Biochem. Biophys. Res. Commun. 123,* 64 (1984)
76. Truneh, A., Albert, F., Goldstein, P., Schmitt-Verhulst, A.-M.: *Nature 313,* 318 (1985)
77. di Padova, F., Schreier, M.: submitted
78. Metcalfe, S.: *Transplantation 38,* 161 (1984)
79. Cohen, S.: in *Biological Response Mediators and Modulators* (ed.) August, J. T., p. 7, Academic Press, New York (1983)
80. Houslay, M. D., Heyworth, C. M.: *Trends Biochem. Sci. 8,* 449 (1983)
81. Kasuga, M., Karlsson, F. A., Kahn, C. R.: *Science 215,* 185 (1981)
82. Mroczkowski, B., Mosig, G., Cohen, S.: *Nature 309,* 270 (1984)
83. Fischer, S., Fargard, R., Gacon, G., Genetet, N., Piau, J. P., Blaineau, C.: *Biochem. Biophys. Res. Commun. 124,* 682 (1984)
84. Hirata, F., Matsuda, K., Notsu, Y., Hattori, T., del Carmine, R.: *Proc. Nat. Acad. Sci. USA 81,* 4717 (1984); Hirata, F.: *J. Biol. Chem. 256,* 7730 (1981)
85. Djiane, J., Houdebine, L.-M., Kelly, P. A., Dusanter-Fourt, I., Teyssot, B., Katoh, M.: Prolactin receptors and intracellular mediator for prolactin action on the mammary cell, in: *Hormones and Cell Regulation* (ed.) Dumont, J. E., Nunez, J., Vol. *8,* p. 269 (1984)
86. Peterkofsky, B., Tomkins, G. M.: *Proc. Nat. Acad. Sci. USA 60,* 222 (1968)
87. Munck, A.: *J. Biol. Chem. 243,* 1039 (1968)
88. Beato, M., Homoki, J., Sekeris, C. E.: *Exp. Cell Res. 53,* 107 (1969)
89. Karin, M., Haslinger, A., Holtgreve, H., Richards, R. I., Krauter, P., Westphal, H. M., Beato, M.: *Nature 308,* 513 (1984)
90. Clark, C. R.: *Trends Biochem. Sci. 9,* 207 (1984)
91. Ryffel, B., Donatsch, P., Götz, U., Tschopp, M.: *Immunology 41,* 913 (1980)
92. Ryffel, B., Götz, U., Heuberger, B.: *J. Immunol. 129,* 1978 (1982)
93. LeGrue, S. J., Friedman, A. W., Kahan, B. D., *J. Immunol. 131,* 712 (1983)
94. Szamel, M., Berger, P., Resch, K.: *Immunobiology 167,* 208 (1984)
95. Schreier, M. H., Iscove, N. N., Tees, R., Aarden, L., von Boehmer, H.: *Immunological Rev. 51,* 315 (1980)
96. Fan, T.-P. D., Lewis, G. P.: *Brit. J. Pharmacol. 81,* 361 (1984); Dewald, B., Geyl, D., Schnyder, J.: unpublished results
97. Palacios, R.: *J. Clin. Immunol. 2,* 155 (1982)
98. Reinherz, E. L., Acuto, O., Fabbi, M., Bensussan, A., Milanese, C., Royer, H. D., Mener, S. C., Schlossman, S. F.: *Immunol. Rev. 81,* 95 (1984)
99. Bendtzen, K., Dinarello, C.: *Scand. J. Immunol. 20,* 43 (1984)
100. Handschumacher, R. E., Harding, M. W., Rice, J., Drugge, R. J., Speicher, D. W.: *Science 226,* 544 (1984)
101. Koponen, M., Loor, F.: *Exp. Cell Res. 149,* 499 (1983)
102. Russell, D. H., Matrisian, L., Kibler, R., Larson, D. F., Poulos, B., Magun, B. E.: *Biochem. Biophys. Res. Commun. 121,* 899 (1984); Russell, D. H., Larson, D. F., Cardon, S. B., Copeland, J. G.: *Molec. Cell. Endocrinol. 35,* 159 (1984)
103. Linzer, D. I. H., Nathans, D., *Proc. Nat. Acad. Sci. USA 81,* 4255 (1984)
104. Lee, L. S., Weinstein, I. B.: *Proc. Nat. Acad. Sci. USA 76,* 5168 (1979); Shoyab, M., DeLarco, J. E., Todaro, G. T.: *Nature 279,* 387 (1979)
105. Hess, A. D., Tutschka, P., Zhang, P., Santos, G. W.: *J. Immunol. 128,* 360 (1982)
106. Reem, G. G., Cook, L. A., Vilcek, J.: *Science 221,* 63 (1983)

107. Orosz, O. B., Roopenian, D. C., Widmer, M. B., Bach, F. H.: *Transplantation 36,* 706 (1984)
108. Krönke, M., Leonard, W. J., Depper, J. M., Arya, S. K., Worg-Staal, F., Gallo, R. O., Waldmann, T. A., Greene, W. O.: *Proc. Nat. Acad. Sci. USA 81,* 5214 (1984)
109. Elliott, G. F., Lin, Y., Mizel, S. B., Bleackley, R. C., Harnish, D. G., Paetkau, V.: *Science 226,* 1439 (1984)
110. Brack, C., Mattaj, I. W., Gautschi, J., Cammisuli, S.: *Exp. Cell Res. 151,* 314 (1984)
111. Horowitz, M., Baron, R., Mart, J., Andreoli, M., Vigrery, A.: *Calcif. Tiss. Int. 36,* 462 (1984)
112. Abb, J., Abb, H.: *Transplant. Proc. 15,* 2380 (1983)
113. Martinet, Y., Pinkston, P., Ozaki, T., Crystal, R.: *Clinical Res. 32,* 244 (1984)
114. Kaufmann, Y., Chang, A. E., Robb, R. J., Rosenberg, S. A.: *J. Immunol. 132,* 3107 (1984)
115. Carpenter, G., Cohen, S.: *Trends Biochem. Sci. 9,* 169 (1984)
116. Granner, D., Andreone, T., Sasaki, K., Beale, E.: *Nature 305,* 549 (1983)

Author Index Volumes 1–3

The volume numbers are printed in italics

MIX
Papier aus verantwortungsvollen Quellen
Paper from responsible sources
FSC® C105338

If you have any concerns about our products,
you can contact us on
ProductSafety@springernature.com

In case Publisher is established outside the EU,
the EU authorized representative is:
Springer Nature Customer Service Center GmbH
Europaplatz 3, 69115 Heidelberg, Germany

Printed by Libri Plureos GmbH
in Hamburg, Germany